# vibe

## An experiential universe

Lance Pollard

Before there were things, there was feeling.

The world is not particles. It is units of experience.

## BEFORE YOU BEGIN

This is a small book about a single idea.

The idea is that the world is made of feeling, and that everything else, matter, space, time, you, is a pattern in it.

I will not ask you to believe this. I will ask you to follow it, step by step, and see what falls out.

The book is built like a staircase. Each short chapter answers one question. Each answer opens the next. You can read it in an afternoon.

Some of what follows is solid. It can be built and run and checked, and it has been. Some of it is implication, a guess about what the same idea means for life, mind, and meaning. I will tell you which is which, in plain words, every time.

You do not need any math. You only need to be willing to picture things.

By the end, you may find you are seeing the same world on a new set of coordinates. Not new facts, just a new floor under the old ones. If even a hint of that lands, the book has done its work.

Everything behind it, the full theory, the math, and the code that runs the world described here, is open here.

github.com/cluesurf/vibe

Go there to see the exact model and run the experiments yourself.

## PART I. THE VIBES

We look for what is real out among the things. The one thing we never doubt is the feeling of being here to experience them.

So start there. With feeling, with experience, vibes, the ground under everything else.

## 1. THE ONE THING YOU CANNOT DOUBT

There is an ache somewhere in your body right now. Or there is rest. Or a small pleasure you had not noticed until these words pointed at it.

Whatever it is, it is happening.

You can doubt almost everything else. The chair might not be holding you. The room might be a dream. The hands you are reading with might belong to a mind that is being fooled about everything.

People have doubted all of this before, on purpose, to see how far doubt can go.

It goes very far. The body can be doubted. The world can be doubted. The past can be doubted, since a memory is only a feeling of having been somewhere.

But the doubting itself is a feeling. And that feeling cannot be doubted away.

To doubt is already to experience something. You cannot get behind it. There is no view from outside your own experience where you could check whether experience is really there.

Try it. Try to step back far enough to ask whether anything is being felt. The asking is felt. You are still inside.

This is a strange kind of certainty. It does not tell you what the world is. It does not even tell you that there is a world. It tells you one thing only, and it tells you that thing completely.

Something is being experienced.

Notice how different this is from everything else you know. Everything else comes to you secondhand. The table reaches you through light and nerves and a long chain of guesses your brain makes. The chain could be wrong at any link.

Experience does not arrive through a chain. It is the place the chain arrives at. It is not reported to you. It is you.

So we have a floor. One solid place to stand. The first and last fact, the one that survives every doubt.

This is unusual for a starting point. Most accounts of the world begin with what is far from us and most uncertain. They begin with particles, or fields, or the vast cold spaces between stars. Then, much later, almost as an afterthought, they try to find a place in that picture for the one who is looking.

We will go the other way. We will begin with the one who is looking, because that is the only thing we were ever given for free.

Everything that follows in this book is an attempt to build outward from this floor. To ask what the world must be like, if the most certain thing in it is that it is felt.

That is a large question and we will take it slowly. For now, stay with the floor itself. Feel how firm it is. Feel that you did not have to argue for it.

The ache, the rest, the small delight. They are not evidence of something deeper. They are the deepest thing you have.

You can lose the world and keep this.

Experience is the one thing you cannot doubt.

## 2. WHAT PHYSICS FORGOT

Imagine a complete blueprint of a house. Every beam, every wire, every pipe. The thickness of the glass. The weight the floor can hold. Nothing is missing from the drawing.

You could build the house from it perfectly. And still, nowhere on the page, is there any mark for what it is like to live there. The warmth of a room in winter. The quiet of it at night.

Physics is that blueprint, drawn for the whole world.

It is an astonishing drawing. It tells you how fast things fall, how light bends, how a magnet pulls, how a star burns and dies. It predicts to a stunning number of decimal places. It has been checked harder than anything else humans have ever made.

And it works by leaving one thing out.

Physics describes the world from the outside. It tells you what a thing does. It tells you how a thing moves, how it pushes, how it responds when you poke it. Every law of physics is, at heart, a rule about behavior.

This is not a flaw. It is the whole method. To measure something, you watch how it acts. You note where it goes and how fast. You never need to ask what it is like to be the thing.

A perfect physics of a person would tell you which nerves fire when they touch a flame. It would trace the signal to the muscle. It would predict the flinch, the cry, the word “ouch”.

It would not contain the pain.

You can describe every step of seeing red. The wavelength of the light. The cells in the eye. The current down the nerve. The pattern in the brain that follows. Lay it all out, leave nothing unsaid, and you still have not said what red looks like from the inside.

There is a gap here. It has a careful name among the people who study it, but the name does not matter. What matters is the size of it.

It is easy to think the gap is a small thing. A last detail to be filled in once the science is finished. One more layer of machinery and the feeling will appear.

But notice what is on each side of the gap. On one side, behavior. Pushes, signals, motions, all the things you can measure. On the other side, the felt inside of the world. The ache from the last chapter. The thing you cannot doubt.

The blueprint has no word for the inside. Not because the inside is unimportant. Because the method was built to look from outside, and what you build to look from outside cannot, by its nature, turn around and face in.

So the gap is not a missing detail. It is the entire inside of the world, left off the page.

This is worth sitting with, because it is so easy to miss. The success of physics is real. The drawing is true as far as it goes. You can be amazed by it and still notice the silence at its center.

We started with the one thing we cannot doubt. The world is felt. Then we picked up the best description of the world ever made and found that it says nothing about the very thing we were most sure of.

Something has to give. Either the felt inside is not really there, which we already saw you cannot believe even while trying. Or the picture is not the whole picture.

We will take the second path. The drawing is good. It is just drawn from one side.

Physics explains behavior. It does not explain experience.

## 3. TURNING THE WORLD OVER

We have two facts that do not fit together.

The world is felt. That is certain. The best picture of the world says nothing about feeling. That is also true.

When two solid facts will not fit, one of your hidden assumptions is wrong. So let us find it.

The hidden assumption is this. We assume matter comes first, and feeling comes later. Stuff is the real thing. Experience is a late trick the stuff learns once it is arranged in a brain.

This is so familiar it feels like common sense. But common sense is just an old habit, and habits can be turned over.

So let us turn it over. Suppose feeling comes first.

Suppose experience is not a trick that matter learns. Suppose it is the substance, and matter is the pattern. Not feeling produced by stuff, but stuff as a shape that feeling takes.

This is a supposition. Hold it lightly. We are not declaring it true. We are asking what the world would have to be, if it were, and whether that world looks like ours.

The move is allowed. Nothing in physics forbids it, because physics never spoke about the inside at all. It described the shapes things make and how the shapes move. It stayed silent on what the shapes are made of. We are free to fill that silence, as long as we change none of the shapes.

So picture water. A wave runs across the surface of the sea. The wave is real. You can see it, ride it, be knocked over by it.

But the wave is not a thing apart from the water. It is a shape the water is holding. The water does not travel with the wave. Each drop bobs in place and passes the motion on. The wave is a pattern moving through a substance that stays behind.

Now do the same with feeling.

Let feeling be the water. Then a wave in it, a stable running shape, is what we have been calling a thing. A stone, a star, a body. Each is a pattern that holds together in the one substance, the way a wave holds together in the sea.

Look at what this buys us. The gap from the last chapter closes. We were trying to get feeling out of matter and could not. Now we do not have to. Feeling was there from the start. Matter is what some of it looks like.

And the gap closes from the right side. The blueprint of physics stays exactly as it was, because the shapes are unchanged. We have only said what the shapes are shapes in.

This is a quiet reversal but it changes the whole direction of the climb. The old way starts with cold particles and tries to reach a mind, and never quite arrives. The new way starts with the one thing we already have, the felt floor, and asks how stable shapes in it could look, from the outside, like a world of solid things.

Now there is a question hiding here, and we should say it before it nags. If everything is a pattern in feeling, what makes one pattern a real thing and another just a passing ripple? A wave is real. A glint of light on the water is barely anything. Where is the line?

That question runs the next stretch of the climb, so we leave it standing for now.

But notice what we have already done. We have stopped treating objects as the bottom layer. A stone is not a chunk of dead stuff with feeling somehow bolted on much later. A stone is a place where the underlying feeling holds a steady, stubborn shape.

Hardness, weight, the way a thing stays put. These are not signs of something other than experience. They are what a very stable pattern in experience does when you push on it.

## 4. WHEN A PATTERN IS REAL

We left a question standing. If everything is a pattern in feeling, which patterns are real, and which are just passing ripples?

Start with a traffic jam. Picture looking down at a packed highway from an overpass.

The traffic has a front edge and a back edge. It has a position on the road. It moves, often backward, against the flow of the very cars that make it. You can point at it. You can be late because of it.

And yet there is no one thing there. No object that is the traffic itself. There are only cars, each one ordinary, each one leaving the back of it as fast as another joins the front. The cars pour through. The shape stays.

So is the traffic a real thing?

Most people feel a pull both ways, and that pull is the interesting part. The traffic is clearly not nothing. It is also not a thing the way a car is a thing. So the question is not whether it is real, but in what way it is real.

We can test it. A pattern earns its reality three ways: by what it can do, by what holds it together while its parts come and go, and by whether it amounts to more than its parts. Take them one at a time, with the wave from before.

It does work on its own. The wave knocks you over, and you do not need to talk about single drops to explain the bruise. The traffic makes you late, and you do not blame any one car. A pattern that has its own effects is not just a name we lay over the parts. It is a player in the world.

It keeps its shape though its parts never stay. This is the one that matters most. The wave holds its form on always-new water. Every drop it had a moment ago is gone, replaced, and it is still the same wave. The traffic holds its form on always-new cars. The pattern lasts precisely because it does not depend on which parts happen to fill it.

And it is more than its parts. You could know everything about one drop and never guess a wave was possible. You could know one car perfectly and never find the traffic in it. The pattern only shows up when the parts are arranged and moving together. Pull them apart and it is simply gone.

Run the traffic past all three. It does work, it makes you late. It holds its shape while cars pour through. It is more than any car. By all three it is real, in the way a pattern is real. Not as solid as a car, but not nothing either. It has its own kind of being.

This gives a ladder instead of a wall. Things are not split into real and unreal. They are more or less stable, more or less able to act, more or less themselves while their parts turn over.

A glint of light on the water fails all three. It does no work, holds no shape, and is nothing without the exact water under it. So it sits at the bottom of the ladder, barely a pattern at all.

A wave passes. Traffic passes. A whirlpool passes, a flame passes. And so, we will find, do stones and stars, and much higher up, living things and minds. Each is a pattern that has climbed far enough to count.

This is the tool we needed. Now we can ask of anything where it sits on the ladder, instead of the blunt question of whether it exists.

And the middle test is the one to carry forward. It is the heart of why anything lasts in a world that never stops moving. Nothing real here is fixed stuff sitting still. Everything real is a shape that survives its own material leaving.

A real pattern keeps its shape while its parts pass through.

## PART II. THE CRYSTAL

One grain of feeling. A net to hold it. Change.

From these three, everything.

## 5. FROM NOTHING

Start with nothing.

Not an empty room. A room is something. Not dark space, not a held breath, not a waiting. All of those are still things. Go under them. Take away the space, the dark, the waiting, the one who waits.

What is left is nothing at all. No here. No there. No before. No after. Not even the difference between full and empty, because difference itself is a kind of something, and we have set everything down.

Sit with that blank for a moment. It is perfectly smooth. Nothing pulls. Nothing points. There is no inside and no outside, because an inside would already be a mark.

Now try to hold it. Just the blank, and nothing else.

You cannot, and why you cannot is where everything begins. To hold the blank is to have it there, before you. The instant there is nothing, there is also the bare fact that there is nothing, and that fact is not itself nothing. It stands apart from the not-being it might have stayed.

And nothing *is*.

That faint standing-there is the first something. Not a thing inside the blank, but the blank caught in the act of being, set against the blank it is no longer. The least possible difference, and it came for free, the moment nothing was anything at all.

This is the smallest something there can be. Not a thing. A difference. A this and a not-this.

Look closely at that one cut, because it is stranger than it seems.

A single difference is never just one piece. It comes with three. There is the one side. There is the other side, its opposite, what the first side is not. And there is the line between them, the edge where one becomes the other.

You cannot have a side without its opposite, or the cut would cut nothing. You cannot have the two sides without the edge that holds them apart, or they would run together back into blank. So the least difference arrives already threefold. A side, its mirror, and the boundary that keeps them two.

Picture a single ripple in still water that has no shore.

Before the ripple, the surface is one unbroken sheet. No place differs from any other. Then, somewhere, the smallest possible stirring. A crest, a trough, and the level line between them. One bump of this against that. The first place where the water is not the same as itself everywhere.

That ripple is the whole event. From a flatness with no features, one feature.

And here is the turn the rest of the book rests on.

A difference cannot hold still.

A crest is higher than the level around it. Being higher, it must fall. The trough is lower, so it must fill. The two sides pull toward each other across the edge, and a pull is already a motion waiting to happen. The instant there is a difference, there is a pull, and a pull cannot just sit there being a pull. It moves.

We did not push the ripple. We did not add a wind. The motion was not laid on top of the difference from outside. It was inside it from the start. To differ is to pull, and to pull is to move.

So nothing, once it is marked off at all, cannot stay nothing and cannot stay still. The first cut tips. The crest spills into the trough. The trough reaches into the crest. One difference becomes a stirring, and a stirring touches the next place, and that place now differs too.

This is the smallest beginning we can find. Not a bang, not a hand, not a law written somewhere first. Just the plainest fact that to mark anything off from nothing is to make a difference, and a difference pulls, and a pull goes.

This is also why nothing could never have stayed nothing. Nothing is a hilltop, not a floor. A floor catches what lands on it and holds it. A hilltop does not. Perfect blankness was never a resting place. It was the steepest peak there could be, and the first difference was already rolling off it.

Everything after this is that single stirring, folding and spreading and never settling. The wave, the stone, the flame, the long climb toward minds. All of it is the first ripple, still moving, because it never had a way to be still.

The world begins with a single difference that cannot hold still.

Here the picture has a hard mathematical spine, and it is worth naming, because from now on the images are only proxies for it. To combine differences and always be able to take them apart again, to lose nothing, is to work in what mathematicians call a normed division algebra. A theorem of Hurwitz says these exist at only four sizes, of dimension one, two, four, and eight, the real numbers, the complex numbers, the quaternions, and the octonions. There are no others, ever. Reversible folding climbs that short ladder and stops at eight, the octonions, the last place where nothing is lost. That single number, eight, and that single object, the octonions, are the seed the whole structure grows from, the same object physics keeps finding under the particles. The pictures are a way in. The octonions are the thing itself.

## 6. A GRAIN OF FEELING

If feeling is the substance, what is the smallest amount of it?

We have been speaking of feeling as if it were like the sea, a single endless body. But the sea is made of drops. A wave is made of countless tiny bobbing pieces. If we want to know what the world is built from, we have to ask about the grain.

So shrink down. Past the wave, past the drop, all the way to the smallest possible piece of experience. Not a thought, not a mood, not even a clear sensation. Those are large and made of many parts. Go below them.

What you reach is a single faint moment of feeling. One pinpoint of being. The least amount of experience there can be, with nothing inside it to take apart further.

Call this grain a *vibe*. A here, a now, a smallest felt point. The grain was the way in, the picture that got us down this far. From here on the word is *vibe*, and grain was only the stepping stone to it.

Picture it as a ripple. Flick the surface of a still pond and a ring spreads from the touch. Pluck a string and a ripple of sound lifts off it. A vibe is the smallest such stirring, one ripple, a single drop's worth of feeling. And there is never just one. The world is a pond struck everywhere at once, countless ripples crossing and blending, each point on the water taking in the ripples that reach it and sending its own back out, on and on. That endless give-and-take is the whole story, and we will come to it. For now, hold the single ripple.

A grain this small cannot hold a picture or a sound or an idea. It is far too simple for that. So what can it hold? The least thing a feeling can carry. A pull, one way or the other. A faint sense of better or worse.

Think of the simplest difference any feeling makes. Before all the detail, before what something is, there is whether it is good or bad. A wince. A sigh of relief. A small wanting more. Strip a feeling down to the bone and that pull is the last thing left.

So a vibe has a tone, the quality of that faint feeling. And there are only a few it can be.

There is the bad pull. A faint pain, a slight away.

There is the good pull. A faint pleasure, a slight toward.

And there is the rest between them. Not bad, not good. A still middle. Call it peace.

Pain, peace, pleasure. These three are the tones a vibe can have. Three values. No more.

Why exactly three, and not two, or many? Because to carry a difference at all you need at least this much. You need a side. You need its mirror, the opposite side. And you need the quiet center they are measured from, the place of no pull.

Two values would give you a flip, on or off, but no center to rest at, no still point. More than three would just be the same axis chopped finer, the same line with extra marks. Three is the smallest set that can hold a real difference and a real rest. One direction, its opposite, and the calm between.

A careful reader will ask the obvious question, so here it is. There are no other numbers down here. A tone is always exactly one of the three, never a four, never a half, never a decimal trailing off. They do not pile up into bigger numbers. Minus one, zero, plus one, the same three at every point, forever.

So the world is discrete. It is built of separate whole steps, not a smooth stuff you can slice forever. And that is a strength, not a lack. A truly continuous thing can be cut in half, and the half in half, with no end, so you never reach a piece to build from. It only ever hands you a smaller piece, never a first one. The bottom keeps receding, which is no bottom at all. A discrete world has a real floor, and you can stand on it.

The smooth numbers we measure the ordinary world with are not lost. They return later, as averages, the way a million coins each falling heads or tails can average to any fraction you like, though no single coin is ever anything but one or the other. Continuity is what the discrete looks like from far off, not what it is made of.

So this is the bottom of the world. Not a particle. Not a number. A single vibe, the smallest grain of feeling, its tone pulling faintly toward pain, or toward pleasure, or resting in peace.

It is almost nothing. That is the point. The smallest thing should be almost nothing, or it was not the smallest. A grain that could hold a melody would still have parts, and we would keep shrinking.

And notice we have not added anything strange to the world. We have not bolted feeling onto matter. We have done the reverse. We have found, at the very bottom, that the smallest piece already feels, faintly, in one of three ways. The feeling was never added. It was there in the grain.

Everything to come is built from these grains, these vibes. The wave, the stone, the flame, the mind. All of them, far up the ladder, are vast arrangements of vibes pulling and resting and passing their pulls along. The whole rich world is this one simple thing, repeated past counting and woven together.

There is another way to hold it, one we will come back to at the very end. Picture the whole as a fire. Each vibe a single flame, and every living thing a flame made of smaller flames.

We have the grain. Next we ask how the grains are joined, because a single grain alone is not yet a world.

But first, fix the grain in mind. It is the seed of all of it.

The smallest piece of reality is a single vibe, with one of three tones.

Stated exactly, a vibe's tone is one of three values, written minus one, zero, and plus one. They are the smallest set that holds both a rest, the zero, and a mirror, a flip that sends plus to minus, what mathematicians call an involution. That mirror is not decoration. It is the seed of spin, the property that later tells matter from force.

## 7. THE WEAVE

A single grain of feeling, alone, is not a world.

It pulls one way, or rests. And then what? By itself, nothing. There is nowhere for the pull to go, nothing for it to touch, no next moment for it to reach. A grain in isolation is a feeling about nothing, going nowhere. It barely is.

So the grains cannot be alone. Each one must be joined to others.

And here we have to be careful, because the habit of mind we turned over in an earlier chapter wants to creep back in. The habit says, first there are the grains, the little things, and then we connect them. First objects, then relations. First the dots, then the lines between.

Reverse it. The thread comes before the knot.

A grain that touches nothing is not really a grain. Its whole nature is the pulling, and a pull has to be a pull toward something. The good and the bad only mean anything in relation. So the grain does not exist first and get joined afterward. It exists by being joined. To be a grain is to be tied to others.

So picture not dots but threads. A weave. A vast net where every crossing is a grain of feeling, and the net is the more basic thing.

Call this net the *mesh*.

The mesh is the fabric of the world. Not stuff. Not particles spread through empty space. A weave of relations, with feeling at every crossing, and the crossings made of nothing but their place in the weave.

This sounds abstract, so make it concrete. Think of a single piece of a jigsaw puzzle. On its own it is an odd little blob, a shape that means nothing. It becomes something, a patch of sky, the edge of a roof, the corner of an eye, only through how it fits the pieces around it. Its whole identity sits in its joins. Lift it out and set it alone on the table and it is no longer anything in particular. The picture lives in the fitting-together, never in any one piece.

The grain is like that. Its pull is a pull toward or away from something, and that is already a relation. A pull with nothing to pull toward is not a fainter pull. It is no pull at all. So the grain needs the rest of the weave, the way the puzzle piece needs the others, simply to be what it is.

Once you see this, the question of what a thing is gets a new answer. We asked it before with the wave and the jam. Now we can say it more plainly. A thing is not a lump in the weave. A thing is a place where the weave holds a steady pattern. A stone is a knot of relations that stays tied. A stable region of the mesh, holding its shape while the feeling at its crossings shifts and passes on.

So we have quietly inverted the usual order of the world. The usual order says, start with separate things, then find the forces between them. We say, start with the relations, the weave itself, and let the things be the stable spots in it. In the old order, the lines are bookkeeping, drawn between the dots that really matter. In ours, the lines are the cloth, and the dots are just where the cloth is tightly tied.

For now, hold the picture, and hold it concretely. Not a heap of separate grains. One living cloth, a vibe at every crossing, each one feeling its neighbors and feeding its own feeling back to them, the whole sheet rippling and settling as one. This is the image to keep close. Call it the vibe mesh, vibing.

One idea is worth meeting before the math arrives, because the whole shape of the world turns on it. Symmetry. A thing is symmetric when you can do something to it and it looks the same after, the way a snowflake turned a sixth of the way around is the snowflake again, or a face flipped left for right is still that face. Gather every move that leaves a thing unchanged and you have its symmetry group, the full list of what you can do to it without changing it. The richer that list, the more balanced the thing.

That word group means something simple. A group is any set of moves you can combine, where doing two in a row gives you another move already in the set, every move can be undone, and one do-nothing move leaves things be. The hours on a clock are a group. Add five hours to ten and you wrap around to three, never leaving the twelve. The turns of a Rubik's cube are a group. When the set is finite, like the clock, it is a finite group, a closed little world of moves that always lands you back inside itself, and those are the easiest to feel.

The simplest move of all is a reflection, a flip across a mirror. And here is the surprising part. Every symmetry there is, every rotation and every fancier turn, can be built by combining plain mirror flips, one after another. So to know all the balanced shapes a space allows, you study its mirrors and the patterns they make as you reflect again and again. Those patterns are reflection groups, and the ones that tile a space are called Coxeter groups, after the man who charted them. They are the bookkeeping of mirrors, and they decide which weaves can exist at all.

There is one more thing to settle before the shape, and it is the setting for all of it. What kind of space does the vibe mesh live in? There are only three kinds it could be, and you can feel each one with your hands.

Lay a sheet of paper flat on a table. That is flat space. Parallel lines stay the same distance forever, the three angles of a triangle add to a straight half-turn, and a circle's rim grows in even step with its width. This is the geometry of the schoolroom, the space your room appears to have.

Now wrap the sheet around a ball. That is closed space. It curves back toward itself, so it holds only so much and then runs out, the way someone walking straight on a globe comes back home. Lines that start out parallel bend together and meet. Triangles puff up fat, their angles adding to more than a straight half-turn. There is less room out here than the flat sheet had.

Now the third, the one with no everyday name, though a Pringle chip, a ruffled kale leaf, and a head of coral all show it. That is open space. It curves away from itself, so it never runs out, and stranger still it gains room faster and faster the farther you go. Lines that start out parallel rush apart. Triangles go thin and starved, their angles adding to less than a straight half-turn. Step out a short way and there is already far more room around you than any flat sheet could hold. This is hyperbolic space.

Those are the only three even spaces there are, the three of constant curvature, named by which way they bend: positive for the ball, zero for the flat sheet, negative for the hyperbolic. Even means the same in every direction at every point, with no special place and no special heading, which is exactly what a world with no favorite spots demands.

The vibe mesh is the open kind, and it has to be. Picture the cloth again, every crossing drinking in its neighbors and feeding its feeling back. A single crossing needs many balanced neighbors, and each of those needs its own many, and each of theirs again, outward without end and without crowding. A flat space fills up too slowly for that, and a closed space runs out altogether. Only open, hyperbolic space keeps handing over more room as you go, room enough for the endless branching the mesh demands. So the rippling cloth lives in the one space that never fills.

How many directions does that space run in, how many dimensions? This too is forced, not chosen. A perfectly regular filling, the same cell repeated with no gaps and no favored places, is a fussy thing to ask for, and in the open curve of hyperbolic space it is fussier still, because the cells must also carry that growing room. The higher the dimension, the fewer arrangements can manage it, and they thin out fast. Climb past four and the regular fillings run out almost entirely. Four is the last floor with a rich enough handful to pick from, so the weave settles there, in four-dimensional open space.

Now the exact shape, which the words net and weave only gesture at. The mesh is a regular tiling of four-dimensional curved space, the hyperbolic honeycomb written {3,4,3,4}, and it grows faster and faster outward because that is what curved space does, a growth those same Coxeter reflection groups track. Its single repeating cell is the 24-cell, the one four-dimensional shape that is its own mirror, with as many corners as faces, built from twenty-four octahedra. Those twenty-four directions are a famous set, the root system called D4, equally the twenty-four unit Hurwitz quaternions, equally the rotations of the binary tetrahedral group. Four roads, one figure.

These perfectly balanced patterns come in a short, special family, and the weave's twenty-four directions are one of them, D4. The largest of the family, E8, is the symmetry of the octonions themselves, the eight-dimensional object the first fold reached. Only a few such perfect shapes exist, so the same ones keep surfacing, here in the weave and again in the deepest physics.

## 8. THE HERE AND NOW

Stand at one crossing of the weave. One single place. A here, a now.

Look around. What is a single spot in the mesh actually like?

It is not a dot. A dot is lonely. A crossing is a meeting. Threads come into it and threads go out of it. It is the place where many relations pass through one point. Think of it less like a pebble and more like a small room.

A room with doors. Each door opens onto a neighbor. Through each door, the weave reaches out to another crossing, and through that door a feeling can come in or go out.

So picture the here and now as a room, and around its walls a set of doors, each leading off in its own direction. Not one door, not two. Many. The room is busy with ways out.

Along each door sits a neighbor, and from that neighbor the room takes in one tone, one vibe's worth of feeling, pulling toward pain, or pleasure, or resting in peace. Each direction brings its own. So a single room is not a single feeling. It is a small gathering of feelings, one along each way the weave runs, all met at one place.

This is what a now is. Not a flat instant. A crossroads. A spot where many directions of the world come together, each bringing its faint charge, all present at once.

Why does it matter that there are many doors and not few? Because the richness of the world has to come from somewhere. If each room had only a door or two, the weave would be thin, a simple line or a flat sheet, and you cannot build much on a line. The many doors give the world room to be complicated. They are why the weave can fold and curve and hold the deep patterns that everything else is made from.

And the doors are not scattered at random. They sit in a particular arrangement around the room, balanced, each one matched by a door facing the opposite way. For every direction out, there is a direction back. The room is symmetric, the doors paired across it like a thing and its mirror.

That balance is going to matter a great deal later. The exact way the doors are arranged is not a free choice and not a detail. It is one of the deepest facts about the world, and there is a whole chapter coming that asks why the doors sit the way they do, and finds that they could sit no other way.

Stated exactly, the room is a node in the mesh, a single junction where the threads cross. And since the mesh is a tiling, the node is one cell of it, one tile. In a flat sheet a cell would be a polygon, a hexagon say, with its neighbors across each edge. In ordinary three-dimensional space it would be a solid, like a die meeting six others face to face. In the world's true four dimensions it is a shape called the 24-cell, with twenty-four faces, and those faces are the doors. That is where the number of directions comes from. Hold the plain room for now. The shape has its own chapter.

For now, just hold the shape of the room. Many doors, paired and balanced, each carrying one faint feeling in or out.

There is one more thing to notice, and it is quiet but important. Nothing about this room is hidden. There is no secret machinery behind the walls, no extra stuff tucked out of sight. The room is its doors and the feelings met along them. That is the whole of it. What you can see at a crossing is all there is at that crossing.

And notice how close the world keeps its business. A room is wired only to the rooms right against it, the ones on the far side of its doors. It feels nothing directly from across the weave, only what its own neighbors hold. This is what is meant by locality. The world acts only where it touches. Anything far off can reach you only by being passed along, neighbor to neighbor to neighbor, never in a single leap. Keep this close, because it is one of the firmest rules there is. Much later we will find a second, hidden sense of near, a back way through the deep that ties far places together. But the plain rule stands first. Each room knows only what is beside it.

So the world, at its base, is many rooms like this, each laced to its neighbors through facing doors, each a small crossroads of feeling. Stand at any one and you are at a now. Step through a door and you are at the next.

This is where everything happens. Whatever the world does, it does by rooms reading their neighbors and settling on a new feeling. The grand patterns, the waves and stones and minds, are all built from this one small act, repeated across room after room.

We have the vibe. We have the weave. We have a single room in it, with its many doors. We do not yet have motion. We have not said what makes a charge go through one door rather than stay. That is the next step, the rule that makes the whole world move.

But the room is the stage, and now it is set.

Every now is a room with many doors.

## 9. THE CELL

Step into one of the rooms and count the doors.

On a tiled floor, each tile touches a few others, edge to edge. A square has four neighbors, a hexagon six. Step up into the solid world and a room becomes a block, a die set among other dice, six broad faces, six rooms next door. Each shared face is a door. But the world is not a floor and not a stack of dice. It is woven in four dimensions, and a room there has more sides than your hands can hold. It has twenty-four.

So picture the smallest dwelling in the world. Not a grain, which is a single point of feeling with no inside to walk around in, but the room that grain sits in, the crossing where its threads meet. That room is one tile of the mesh, shaped like nothing you could set on a table. It has twenty-four flat faces, and every face is a door onto a neighbor. Twenty-four ways out, twenty-four rooms pressed against this one, each sharing a wall.

This shape has a name. It is the 24-cell, and it is the most balanced object its dimension allows. Turn an ordinary die and its corners go one place, its faces another, sharp points and flat planes, plainly different things. The 24-cell has as many corners as faces, twenty-four of each, so it is its own mirror. Swap every corner for a face and you get the very same shape back, unchanged. Nothing in it is favored. No door is wider, no corner sharper. It is a crystal so even that it looks the same turned inside out as right side in.

The twenty-four doors are not scattered at random. They come in twelve facing pairs, each door squarely across from its partner, the way north faces south. That evenness is why so much will later hang on this one shape. A facing pair is what lets a thing turn and still know which way it turned. The balance is what keeps the world from leaning toward any one direction. Spin, charge, the gap between a thing and its reflection, all of it will be read off these twenty-four doors.

So the world climbs in three plain rungs. A grain is a single feeling, a lone point. A cell is the room that grain keeps, with its twenty-four doors. The mesh is every room at once, clasped door to door to door, outward without end. Grain, cell, mesh. A point, a room, a city with no edge.

Hold that picture. Why a room must take this shape, and no other, is the next chapter.

Stated exactly, the cell is the 24-cell, the one regular four-dimensional solid that is its own dual, with twenty-four corners and twenty-four octahedral faces. Its twenty-four directions are the root system D4, the same as the twenty-four unit Hurwitz quaternions, and the same as the rotations of the binary tetrahedral group. They are the tightest perfectly balanced set of directions that four-dimensional space allows, which is the deep reason the whole world is built on them.

## 10. THE ONLY WEAVE

We have a world. Grains of feeling woven into a mesh, each crossing a place where many threads meet, all of it pulling and passing the pulls along.

But we kept saying the shape of that crossing is not a free choice, and kept promising the reason. Here it is.

Of all the ways a weave could be made, why this one?

Go back to the first stirring. A single difference that could not hold still. It pulled, and the pull reached the next place. So it folded onto itself. The crest met the trough, and where they met, a new difference. That folded too. Each fold was forced by the one before, the way one ripple makes the next.

Folding is just combining. Two things meeting to make a third. And there are only so many ways things can combine.

Picture a short ladder.

The lowest rung is the plainest joining. Things line up, and it does not matter which you take first. Simple, calm, a little poor. Climb a rung and order begins to matter. Take this then that, and you do not land where you would, taking that then this. The world has gained a before and after.

Climb again. Now turning matters. To combine two things you must mind which way each one faces, because a turn one way and a turn the other do not cancel cleanly. The joining has grown a handedness.

Climb once more, to the last rung. Here the joining is richest and strangest. So much happens in a single combining that you can no longer even group the steps freely. Bundle them one way or another and you reach different ends. It is dizzying and full, the most a combining can carry.

And then the ladder stops. Try to climb higher and the thing falls apart in your hands, too loose to hold together at all. The top rung is the richest joining there can be. Beyond it, nothing more can be added.

So the folding does not wander forever. It climbs this short ladder, forced at each step, and locks at the top. One special crossing built from it, a room with many balanced doors, the most a single place in the weave can hold.

That is what fixes the shape. Not taste. The end of a ladder that has nowhere left to go.

Those four rungs have exact names, and the ladder really is only four rungs high. They are the only number systems in which you can always divide, and they climb by doubling in size. First the ordinary numbers, the plain line we count on. Then the complex numbers, which add a turn, a way to spin in a plane. Then the quaternions, which add handedness, where the order of two turns changes where you end up. And last the octonions, where even free grouping is gone, where bundling the same steps differently lands you somewhere else. Their sizes run one, two, four, eight, doubling each time. And then it stops. A theorem proves there is no fifth, that octonions are the last place division survives. Eight is the top rung. This is why the octonions, the eight-part numbers, keep turning up beneath everything in this book, and why the largest of the perfect balanced patterns, the one called E8, is exactly their symmetry. The first fold of the world climbs this little ladder and halts at eight, because there is no ninth thing to become.

But being richest is not enough. The crossing must also build a livable world. So lay down what the weave must do, and pour every imaginable net through a sieve.

The first demand. It must fit together perfectly. The same crossing everywhere, no gaps, no seams, no favorite places. A world cannot have special spots. Most imaginable nets cannot tile like this at all. They fall.

The second demand. It must curve the open way, the curve that leaves a crossing more room around it than a flat sheet would, room to branch and grow. Many survivors cannot hold that curve. They fall.

The third demand. It must carry a twist. There must be a way for something to turn inside it and come back changed, so the world can tell a turn from its mirror, a thing from its reflection. This is the deepest demand, and almost nothing meets it. Most of the few left have no place to keep a twist. They fall, nearly all of them.

The fourth demand. However wild it is up close, stepping far back, its far edge must smooth into the plain, roomy space we live in. The strange deep weave must look, from a distance, like the ordinary openness around you now. Almost nothing that came this far can do that too.

Four demands, each simple, together merciless. Pour the whole sea of weaves through and watch it drain.

When the draining stops, a single weave is left in the sieve.

Just one. Its crossings have the doors they have, balanced as they are, because no other arrangement survives all four at once. And the decider, in the end, is the twist. You can find plenty of weaves that tile and curve and smooth out nicely, but they are dead. They carry no twist, so they hold no matter, so nothing could ever live in them. The survivor is the one weave that can.

So we did not draw a clever shape and admire how well it worked. We wrote down a few plain things any livable world must do. The shape fell out as the only thing that could do them all.

The shape is not chosen. It is the only one left standing.

Here are the names behind the sieve. Which curved tilings can exist at all is settled by Coxeter reflection groups, the algebra of mirrors, and of the four-dimensional hyperbolic honeycombs they allow, only the one written {3,4,3,4} passes all four demands. The last demand, that the wild deep weave look flat and roomy from far back, is met on a horosphere, a sphere grown so large that its skin is perfectly flat. The plain three-dimensional space you stand in is that flat skin, carried inside the curved four-dimensional bulk.

## 11. ONE LAW

We have rooms and doors and charges. No beat has fallen yet. The world is set up but nothing is happening.

What makes it go?

You might expect a thick book of laws. Physics has many equations, after all, one for gravity, one for light, one for each kind of thing. So you might brace for a long list.

There is no list. There is one rule, and that is it. The same rule, in every room, at every step, forever.

Call it the *knit*, because it is what ties the next moment to this one.

Here is the one move, and it is not the one you would guess. A room does not send anything anywhere. Nothing leaves it. Nothing crosses to it. The room takes in the tones its neighbors are holding, the ones it sees through its doors, and blends them, with its own, by a fixed pattern, into its next tone. That is the whole rule. Take in the neighbors, settle on the next feeling. Every room does it at once, and the world has stepped once.

Be stubborn about this, because the wrong picture comes so easily. There are no little charges traveling the halls, slipping out one door and in the next. The rooms are the space itself, and space has nowhere to go. The only thing that ever changes is the feeling each room is holding. Nothing travels. The pattern of feeling is made again, room by room, and that remaking is the whole engine.

One such round, every room reading its neighbors and settling on its next charge, is a beat. The world steps one beat, then another, then another, the same reading done everywhere at once.

Then how does anything move? The way a wave moves. Picture a stadium crowd doing the wave. No one runs around the ring. Each person only rises and sits, watching the people beside them. Yet the wave travels all the way around, real, with a speed, made of no one who travels. Everything in this world moves like that. A shape of feeling, made again one room over each beat, while every room stays home. The shape travels. The rooms never do. A ripple, a stone, a body, a thought, each is a pattern that moves by being remade just to the side of where it was.

That is it. The waves, the stones, the stars, the living things, the minds. All of them are this one little step, every room taking in its neighbors, repeated past counting across the weave.

It seems far too simple to carry a world. But that is how the deepest things work. The same small reading, done everywhere, again and again, can build any amount of richness. Complexity does not need a complicated rule. It needs a simple rule and enormous patience.

Now, two things about this rule are easy to miss, and they are the most important things about it.

The first. The rule loses nothing. When the rooms settle on their new tones, none is made and none destroyed. The feeling is only remixed, the same total held exactly, room to room and beat to beat. The books always balance.

The second follows from the first. Because nothing is lost, the rule can be run backward. Take any moment and you could, in principle, undo the last beat exactly, and the one before that, and so on, all the way back. No information is ever erased, so no step is ever truly gone. The path can always be retraced.

This is a strange and powerful property. The world it describes never forgets. Every beat is fully recoverable from the next. The whole history is, in a sense, still there, folded into the present, never thrown away.

Sit with how unusual that is. Most of what we see in life seems to forget. A cup shatters and does not gather itself back. Heat spreads and does not return. We will have to face that later, because the bare rule says forgetting should be impossible, and yet the world we live in clearly forgets. There is a real puzzle there, and a real answer, but it waits.

For now, hold the rule itself. One step, every room taking in its neighbors and settling on its next feeling. Repeated everywhere, forever. Never losing a thing. Always able to run backward.

A rule like that is not a command barked at the world from outside. It is just what the weave does, what it has always done, the way it cannot help but behave. A regularity so deep and so unbroken that we call it a law.

The exact name for the one law is a reversible cellular automaton. Each room's next charge is fixed by its own and by the ones it reads in its neighbors, one local rule run everywhere at once. It is a reversible permutation in exact whole-number arithmetic, so it conserves its total and can be run backward to recover any past. Of the thousands of rules the cell's symmetry would allow, exactly one survives every demand.

## 12. THE GROWING EDGE

We have a rule that runs the world, beat by beat. But we have been quietly assuming something. We have been assuming there is a next beat for it to run into. A place for the world to go.

Where does that come from? Where does time come from?

The easy answer is that time is a backdrop. A clock hung outside the world, ticking on its own, and the world plays out against it. The rooms and charges do their dance, and time is the empty stage they dance on.

Let us not take the easy answer. A clock outside the world is one more thing we would have to explain, and we have been trying to build the world from the inside, from feeling and weave and rule, with nothing bolted on from outside.

So picture instead a crystal forming in still water. You can think of it as a jewel, a beautiful stone, slowly taking shape.

A crystal does not appear all at once. It grows. There is a settled body of it, already formed, locked into its pattern, finished. And there is an edge, where new pieces are still joining, still finding their place. Past the edge there is only the open water, not yet crystal at all.

The weave is like that. It is not a finished thing sitting in time. It is a crystal still growing.

And there is no outside to it. The crystal is not sitting in some larger room. It is everything there is, the whole of what exists, the all. The open water past the edge is not another place. It is only this same crystal's own not-yet, the part that has not grown into being. If you want to picture other worlds, or many universes, they are not separate crystals set beside this one. They are regions within it, walled-off quarters of the one weave, sealed enough to feel apart.

There is one crystal, one vibe mesh, and all of it is inside.

Look at the three regions.

There is the settled inside. The rooms whose beats are done, whose charges have met and moved on, locked now into the pattern they made. Nothing there will change again. This is the past. Not gone, not erased, remember the rule loses nothing, but finished. Set.

There is the edge. The rooms forming right now, where charges are meeting and about to move, where the pattern is being decided this very beat. This is the present. The thin live skin of the crystal, the only place where anything is actually happening.

And there is the open water beyond. The rooms not yet formed. The beats not yet taken. This is the future. Not a place that exists and waits to be reached. A place that does not exist yet, that the growing edge is about to make.

Now see what this does to time. Time is not a stage the world sits on. Time is the growing itself. The crystal does not move through time. The crystal getting bigger is what time is. Each new layer at the edge is another moment, made by the making.

This is why time has a direction, and why the direction is not a separate law we had to add. The crystal grows outward, into the open water, never inward into the finished part. The settled inside is done and stays done. The edge only ever advances. There is no growing backward, because the inside is already laid.

So the arrow of time, the reason it runs one way and not the other, is just the direction the crystal grows. Past behind, future ahead, present the bright edge between. Not a rule about time. The shape of the growing.

Notice how this fits with the rule from before. The rule could run backward, we said. Every beat is recoverable. And yet time runs forward. There is no contradiction. The rule says the finished crystal could in principle be read in either direction, like a book you can read back to front. The growing says the book is still being written, and the writing only goes one way, page after page, the edge always the latest line.

So we never needed a clock. The world makes its own time by making more of itself. Every present is the edge of a thing still under construction. Every past is a layer already set. Every future is open water the edge has not reached.

Time is not running out. Time is being made. The world is not a film already shot, playing toward an ending. It is a crystal still forming, a coastline still being drawn, larger at every beat.

You are at the edge right now. Reading this is the crystal growing. The next moment is open water, not yet made, about to be.

Time is the world still being built.

The exact statement is the one physics calls the arrow of time, the second law. The local rule carries no direction, it runs as cleanly backward as forward. The direction lives only in the growth, the steady adding of new cells at the boundary, a count that can only rise.

That one-way rise is the arrow, and the same growth is what physics measures as the expansion of the universe.

## PART III. THE LAWS

Step back far enough and the pixels become a picture.

The laws are not added from above. They are what the vibes look like from far away.

## 13. FROM NUMBERS TO FIELDS

Hold a photograph close to your eye and you see dots. Step back across the room and the dots are gone. A face is there instead, smooth and whole.

Nothing in the photo changed. Only your distance did.

The world is built the same way. Up close it is vibes. Each vibe carries a small whole number, a charge that is one of a few plain values. There is nothing smooth about it. Counting is all there is.

Smooth physics does not live at that scale. It lives at the scale of the room.

To go there, you blur. You stop tracking each vibe and start tracking averages over little patches of the weave. How much charge sits in this patch. How fast it is pulling toward the next. The whole numbers melt into a flowing quantity that can take any value, the way a thousand separate dots melt into one gray.

This blurred quantity is a field. It is not a new thing added to the vibes. It is the vibes, seen from across the room.

Now ask what the field does. To answer, look back at the one law underneath.

Each room only takes in the tones it sees and settles on its own. It mixes them, it never adds to them. It never reaches in and makes a new charge from nothing, and it never quietly deletes one. Whatever charge enters a patch had to come from a neighbor. Whatever leaves had to go to a neighbor.

This is the one thing the law refuses to do. It will not create or destroy charge.

That refusal is small and exact when you are counting vibes. It is just bookkeeping. Nothing in equals nothing lost.

Blur it, and the bookkeeping becomes a law of flow.

If a patch holds less charge than it did a beat ago, the missing amount did not vanish. It crossed the patch's edge. So the change inside a region is always equal to the flow through its boundary. A drop in here is a current out there. The two can never disagree.

Physics already has this law. It is called a conservation law, the rule that water, electricity, and charge all obey: the stuff is neither born nor lost, only carried from place to place. Write it for the blurred field and you get it exactly, with no slack and no fudge.

So the smooth law was hiding in that refusal all along. Down in the vibes, the rule will not make a charge or destroy one. That exact refusal to gain or lose anything is what the word conservation means. The flow law is the same refusal, seen from far back.

This one is firm. Not a hope, not a near miss. The vibes hold their count by plain arithmetic, checked one by one, whole numbers that must balance to the last unit. Blur them and the smooth law comes out exact, the same on a flat weave and on the curved weave the world is actually built from.

That kind of firmness is rare in this book. Most of what comes later only runs alongside known physics, reaching the same answer by its own road, and some of it stays open. This does not. This one you can watch the vibes build.

Keep the picture. Dots become a face, and counting becomes flow. The one thing the rule never touches is the total charge, and that is exactly what physics says can never be made or destroyed.

Nothing made, nothing lost. Blurred, that one fact is the law of flow.

The exact framework is quantum field theory, the language modern physics writes matter and force in. A field is a value carried at every point, and the smooth fields of physics are what the discrete values look like averaged over many cells. Conservation, blurred into a smooth law, is a continuity equation, the backbone of every field theory.

## 14. SPIN

The smallest pieces of matter carry a strange habit of turning. Turn a thing all the way around, a full circle, and you expect it back the way it started. The electron is not. It takes two full turns to come home. Physics has known this for a century and never had a plain reason for it. The weave does, and it falls out of the shape we have already built. Start with a trick you can do with your own arm.

Hold a cup of coffee flat on your palm. Now turn it under your arm and back over the top without spilling. One full turn leaves your arm twisted. You have to go around again, a second full turn, before your arm hangs free.

Two turns to come home. One is not enough.

Belt dancers know the same trick. Give a belt a flip and a full turn does not undo it. The second turn does.

This is not a quirk of arms or belts. It is a feature of how turning works in a connected world. Some things, once turned all the way around, are not the way they started. They are flipped. Only a second full turn restores them.

The weave has this built into it.

Recall that every room in the weave opens onto a fixed set of directions. Those directions are not a loose bag. They lock together in a tight pattern, each one bound to the others. That binding is what makes turning mean something. You cannot rotate one direction without the whole set responding.

And the way they respond carries the twist. Turn the set of directions once, all the way around, and certain patterns riding on them do not return. They come back flipped, like the arm under the cup. Turn again and they settle.

A pattern that needs two turns to come home is the simplest deep object the weave allows. It is not put in by hand. It falls out of how the directions are wired.

These two-turn patterns are exactly the kind of thing the electron is.

The electron has long been a puzzle in this one respect. It carries a turning that a single rotation does not undo. For a hundred years this was just a fact you wrote down and used. It worked, but no one could say why a thing should need two turns rather than one.

In the weave it is not a separate fact. The two-turn behavior is the same behavior the directions already have. The electron and its near relatives are the simplest twists the weave can hold. Their strange turning is the weave's turning, shown at the smallest scale.

This much is shown. The two-turn twist is there in the structure, not added to fit the answer. The directions carry it on their own.

Something else comes along with it. When you sort the twists the weave allows, they do not arrive as one kind. They arrive in families, repeats of the same shape at heavier and heavier weights. Physics sees this too. The electron has two heavier cousins, alike in every way but mass, and the other matter particles copy the pattern. Three families, and no more.

The weave produces families. That part lines up. But the plain count, why exactly three and not two or four, does not yet drop out cleanly. It may be hiding in the structure. It may need something not yet seen. For now it is open, and calling it shown would be claiming more than is there.

So keep the two things apart. The twist itself is earned. The number of families is a question still standing.

Stay with the cup. The arm that will not come home in one turn. The world is woven so that some of its smallest pieces are like that arm. They carry a memory of having been turned, and only a second turn clears it.

Now the exact math, which the belt trick only mimics. A thing that must turn twice to return is a spinor, and spinors live in a Clifford algebra, here the very algebra built by multiplying the octonions, the eight-part numbers from the first fold whose unit directions are the building blocks of the 24-cell, the single cell that tiles the whole mesh. The twenty-four directions form what is called the double cover of rotation, the group SU(2), short for the special unitary group of degree two, in which one full turn counts as minus one and only the second brings a thing home. Run the rule and a single clean matter particle moves with the spectrum physics calls the Dirac equation, carrying a magnetic strength of exactly two, the measured value, put in by nothing.

## 15. THE FORCES

Nature runs on a handful of forces that look nothing alike. One glues atoms together, one pulls things to the ground, one drives the slow fire of the sun. This chapter takes up a long-standing idea, that they are not separate at all but one thing broken into pieces, and asks what the weave adds to it. Begin with light through glass.

Shine white light through a glass prism and it fans out into a band of color. The colors were always in the light, folded together. The prism only unfolds them.

The forces of nature look like that band. From far away they seem separate. One holds atoms together. One makes things fall. One runs the slow burning of the sun. Up close, the picture says they are one light, split.

Start with the weave's deep pattern. The directions in each room bind together into a single tight symmetry, one balanced arrangement that the whole structure respects. It is not several rules stacked. It is one shape, very full, holding many ways of turning at once.

A symmetry that full does not stay folded. As the world settles and cools, it breaks. The one shape comes apart into smaller pieces, each piece a way of turning that survives on its own.

Those surviving pieces are the forces.

This is the prism, run in reverse of how we usually meet it. We meet the colors first, the separate forces, and only later guess at the white light behind them. The weave offers the white light first. The colors fall out when it breaks.

The idea that the forces are one broken symmetry is not new to the weave. Physicists have chased it for decades and built strong versions of it. What the weave adds is a reason the symmetry has the shape it does. The shape is not chosen to fit the forces. It is the same shape that ties the directions together, the same shape that carries the twist of the last chapter.

Some of what comes out of this is genuinely earned. One of the numbers that describes how the forces mix, a particular angle that says how two of them blend, comes out of the structure at a clean value close to what is measured. That is a real result, drawn from the shape and not slipped in.

Not every number here is derived. To compare the weave with the known particles, several of the measured quantities are fed in by hand. The masses, the strengths, the fine settings that the Standard Model lists, many of these are read from experiment and placed into the picture, not produced by it.

When a number is fed in, the weave is matching, not predicting. It is showing that its structure can carry the known values, which is worth something. It is not showing that it forces them. Those are different claims, and only the first is true here.

So mark each one. The breaking into forces, shown in spirit and along a road others have walked. The mixing angle, a real derived value. The masses and strengths, put in by hand for now, running parallel to known physics rather than falling out of the weave.

It would be easy to blur these together and let the earned result lend its shine to the fed-in ones. That blur is exactly what the prism warns against. Keep the colors named.

Stay with the light. One full symmetry in the weave, too rich to hold itself together. As the world cools it splits, and the splinters are the forces we know. Some of what splits off we can predict. Some we can still only match.

The forces are one symmetry, broken.

The exact name is gauge symmetry, and the exact starting group is the one physicists write SO(10), short for the special orthogonal group of degree ten, a grand unified symmetry that breaks into the three forces of the Standard Model, its weak part the same SU(2), the spin group from the last chapter. The single breaking even fixes the number called the weak mixing angle, which runs to its measured value.

## 16. LIGHT AND THE LIMIT

The world has a top speed, the speed of light, and everyone measures the same value for it no matter how they are moving. That has puzzled people for a long time. In the weave it stops being strange, because the limit is simply how fast the fabric can pass a change along. Begin with a stone and a pond.

Drop a stone in a still pond. A ring spreads out. The ring is not a thing moving through the water. It is the water passing a disturbance along, neighbor to neighbor, each patch nudging the next.

The ring cannot outrun the water it rides on. There is no faster way for the news of the stone to travel than from one patch to the one beside it.

The weave carries change the same way. A pattern in one room can only show up in the next when that neighbor reads it, one beat later. Nothing is carried across. Each room only takes its next feeling from what sits beside it. That is the only motion there is.

So nothing in the world moves more than one room per beat. Not because a sign forbids it. Because there is no other way to go. A change has to be passed hand to hand, and the fastest hand-off is to the room next door.

That cap is the speed of light.

Light is not special for being fast. Light is what travels at the only top speed the weave allows. Anything that moves at all moves by passing along the weave, and the swiftest passing is one room per beat. The cap is not a rule about light. It is a rule about the weave, and light simply rides it at full pace.

This already explains why the speed of light is the same for everyone, a fact that puzzled people for a long time. It is not a property of the lamp or the eye. It is the refresh rate of the fabric itself. Every observer is made of the same fabric, so every observer measures the same top speed.

From that one cap, more follows.

If signals can only travel so fast, then a thing in motion cannot keep time the way a thing at rest does. Its inner workings depend on changes passing back and forth inside it, and those changes are already spending part of their budget on the motion. So the moving thing ticks slower. Its clock falls behind.

This slowing of time with motion is a known effect, tested countless times. In the weave it is not an extra law. It is what you get once you accept that nothing beats one room per beat. The cap leans on everything, and time bends to fit it.

So far this runs alongside known physics by a clear road. The top speed, the shared value, the slowing clock, all of it reads off the weave's one step.

The fuller story is harder, and this is where we should slow down and be careful.

The full theory of motion and time, the one that says how space and time trade off in every direction at once, has a property that sits awkwardly on a fixed weave. It treats all smooth turns of space and time alike, including ones that should stretch and pull the very grid the world is built on. A fixed grid has a built-in frame. Getting the full smoothness to emerge from it, with nothing left over that would let you feel the grid underneath, is not yet done.

This is the sharpest open problem in the whole picture. The speed limit is shown. The slowing clock follows. The complete theory of motion on a fixed weave is still being worked out, and pretending otherwise would be dressing a hope as a result.

Stay with the pond. The ring that cannot outrun its water. The world has a ring speed, and we call it light.

The speed limit is just how fast the weave can pass a change along.

The exact name for the limit is Lorentz invariance, the symmetry of Einstein's relativity, and here it emerges rather than being assumed. The one-cell-per-beat ceiling is the light cone, and the spacetime symmetry built on it is Lorentz, with any violation pushed below sixth order, far under what the sharpest test, the timing of gamma-ray bursts across the sky, could ever see. The control matters, on a flat lattice the same construction fails.

## 17. GRAVITY

This chapter asks what gravity is, and underneath it, what space is. The weave's answer is that distance is not a stage things sit in but a count of how tightly two regions are linked together, and gravity is that linking drawn denser in one place than another. Begin with a floor of tiles.

Picture a floor of square tiles, except there is no floor underneath them. Each tile is held beside its neighbors only by small clasps along its edges, and those clasps are the whole of it. There is no ground keeping the tiles in place. The clasps are what make them one connected surface at all.

Now ask how far apart two tiles are. You count the steps. From a tile to the one clasped beside it is one step, to its neighbor another, and the distance between two far-apart tiles is just how many you cross to walk from one to the other. The distance lives in the clasps, in the chain of connections, not in any space the tiles sit in.

Now unclasp two patches of the floor completely. Cut every connection between them. They do not drift apart, and they are not far. There is simply no number of steps from one to the other anymore, no between at all. They have become two separate surfaces, the way two floors in two different cities have no distance you could pace out tile by tile.

Space is held together like that.

Two regions of the world are not side by side because they sit in some larger box. They are side by side because they share many links, connections in the weave that tie one region's tones to the other's. The more links they share, the closer they are. The fewer, the farther.

Distance is a count of shared links.

This sounds strange until you notice what it explains. If nearness is just heavy linking, then to be far is to be loosely linked, and to be infinitely far is to share no links at all. Geometry stops being a stage the world sits on. It becomes a record of how the world is tied to itself.

The links have a name in physics. The shared connections between two regions are what is meant by entanglement, the deep togetherness that makes two parts of the world answer as one.

So watch what happens when you change the linking. Add links between two regions and they draw closer. Cut links and they drift apart. If you could strip every shared link from a patch, the space inside it would not shrink. It would come undone. There would be no inside left to measure.

Now a stranger thing about the links.

The weave keeps a complete record. Every meeting of tones, every settling, leaves a mark, and the mark is never finally lost. The weave does not forget. What happened is linked in.

Now something that should stop you. To read the whole inside of a region, you do not have to go inside. You can read it from the edge. The boundary that wraps a region holds, in its links, everything within. The way a whole scene can be pressed onto a flat plate and read back from the plate alone, the inside of a piece of the world is written on its skin.

It is a quiet, exact bookkeeping. Nothing falls through. The ledger always balances, and the ledger is the surface.

Now stand back and ask where these links actually run. Not across the flat space you see. The space you live in is only a skin, a thin flat surface wrapped on a deeper, four-dimensional bulk underneath, the way the rind of an orange wraps its pulp, or the leather of a baseball wraps its wound core. The links dive off the skin and run down through that bulk.

And the bulk is small. Strangely, wonderfully small. Down in the deep interior every point is only a few steps from every other, so two places that sit a world apart on the skin can be almost touching underneath. Picture two trees far apart in a field. Up in the air their branches never come near, but their roots reach down and tangle in the same dark soil. Or crumple a flat sheet of paper in your fist, and two corners that were far apart along the page end up pressed together. Far along the surface, near through the deep.

This is the hidden sense of near that the room had promised, back when we said each place touches only its neighbors. The same two places can be wired close underneath, through the bulk, and that secret closeness is what we have been calling a shared link. Entanglement, the deep togetherness of two far-apart things, is two points on the skin whose roots nearly meet below.

There is a way to feel which side is in charge. Think of a puppet. The puppet is the flat skin, the figure you watch. The strings rise from above, and the hand that works them is the bulk. The figure dances, and every motion of it traces to a pull from the deeper place you do not see. The four-dimensional bulk holds the strings. The three-dimensional world is the puppet they move.

One more turn, and it is the subtle one. Because the skin is a flattened shadow of the bulk, a single arrangement down below can show on the surface in more than one way. The same deep nearness might surface here as two things plainly close, and there as two things far apart yet entangled, depending on how the bulk folds up to meet the skin. The fact underneath is one fact. Its faces on the surface are many.

Gravity is the pull of dense linking. Where the weave is linked tighter, the measure of space is drawn in, and things drift toward the tighter linking the way a ball rolls toward the low point of a dip. Clocks slow there, paths curve, the geometry sags. Gravity is the weave being drawn tighter in one place than another.

There is a known and powerful way to make this exact. Treat the shared links the way you treat heat and pressure, as something with a tally that wants to balance, and the bending of space comes out as the natural settling of that tally. That argument needs the world to have a finite count of states in any region. A smooth space has no such count. The weave does. It is vibes, and vibes can be counted. So the finite tally is there for free.

Some pulls run so deep the light from beyond cannot climb back to you. That edge is a horizon. It is not a wall in the world. The things past it are still there, still linked into the weave, only their links no longer reach your eye. A horizon marks the limit of what can be known from where you stand, not the end of what is.

Mark where this stops. The reading of gravity as linking runs parallel to a known and strong argument. The finite count is supplied by the vibes. The complete field law, the exact strength of the pull, does not yet fall out of the weave alone. That is still open.

Stay with the tiles. Two patches, linked. Cut every link between them and the space between them is not far. It is simply not there.

The exact statement is Einstein's, the field equations of general relativity, recovered in the flat region as the law of how the weave's curvature answers to what sits in it. Because the cell sets a smallest length, curvature can grow large but never infinite, so the singularity at the heart of a black hole, or at the beginning, is softened to something finite.

## 18. THE QUANTUM

The quantum world is usually told as a world of real chance, with nothing settled until you look. The weave tells it without any dice, as a settled world seen through partial knowledge, and is up front about the one piece that remains open. Begin with a spinning coin.

Spin a coin on a table. While it spins, ask which face is up. There is no answer yet. But the moment it falls and rests, it is heads or tails, and it was always going to be that one. The spinning only hid it from you.

The quantum world is often told as a world of dice. The pieces, the story goes, truly have no answer until you look, and the looking forces a real roll of chance.

The weave tells it without dice.

The law that moves the weave never flips a coin. Every beat follows from the last by a fixed step. Run the world again from the same start and you get the same result, down to the last vibe. There is no randomness anywhere in the base. It is settled all the way through.

And yet, from inside, it looks like chance.

The reason is simple once you see it. You are a small pattern in a vast weave. You can never hold the full state of everything around you. You see your own corner and guess at the rest. Your knowledge has a horizon, and most of the world lies past it.

When you cannot see the whole settled state, the part you can see behaves as if it were undecided. Not because it is. Because the deciding factors are out past your horizon, like the spinning coin's hidden face.

So apparent chance is not a roll. It is a settled world viewed through partial knowledge.

This is more than a comforting picture. The weave reaches the actual numbers. The pattern of quantum odds that experiments measure, the rule that says how likely each outcome is, comes out of the weave. So does the exact limit on how tightly two distant parts can agree, a famous bound that no dice-world can fake and that the weave meets on the nose. These are real results, running alongside known quantum physics by a different road.

Now look at what measuring is.

A measurement is not a magic act of looking. It is a small system touching a huge one. A single vibe-pattern brushes against a vast surround, a detector, an eye, a room full of air. The small thing was poised between possibilities. As it couples to the enormous surround, it settles, the way a spun coin settles when it finally meets the table.

Settling is relaxing into one state under the weight of everything around it. The surround is too large and too tangled to ever spin back. So the small system commits. What looked like a choice was the world coming to rest.

This part is shown in its mechanism. No coin is flipped. The settling is just the law running where a tiny thing meets a vast one.

One question stays open, and it is the hard one. The weave gives the spread of outcomes and their odds. It does not yet say why, on this run, you find this single outcome rather than the whole spread. The odds are derived. The selection of one particular face, in one particular world, is not yet pinned down. That wall still stands, and naming it solved would be claiming more than is there.

So hold the two apart. The odds and the agreement limit, reached by a real road. The settling of a measurement, shown as mechanism. The choice of one outcome over the others, still open.

Stay with the coin. Already heads or tails the instant it lands, hidden only while it spun. The world is like that everywhere, settled beneath what any one corner of it can see.

The exact objects are the Born rule, which turns the size of a pattern into a probability, and the Bell test, whose score this local, deterministic base drives all the way to two times the square root of two, the Tsirelson bound, the ceiling quantum mechanics allows and no classical story can reach. A control gives the lower, classical value, so the number is measured, not assumed.

## 19. THE EXPANDING WORLD

Watch a coastline being drawn on a map, slowly, by a hand that has not finished. The drawn part is firm. The pen sits at the working edge. Past the pen there is only blank paper, not yet land.

The map is growing. New coast appears at the edge, beat by beat, where the pen still moves.

The weave grows like that.

The settled inside of the weave is the past, finished and quiet. The working edge is the present, where new rooms are being made. Out past the edge there is no thin space waiting to be filled. There is nothing yet. The edge is where there starts to be a world at all.

So the world is not a fixed box with stuff rattling around in it. It is a fabric still being woven, larger every beat because new threads are added at the rim.

This gives the expansion of space for free.

Astronomers see the universe growing. Distant galaxies drift apart, and the farther ones drift faster, as if the space between everything were swelling. For a long time this was a fact to accept and describe, a stretching with no clear reason underneath.

In the weave the reason is the growth. The world expands because new rooms keep being made at its edge. Space gets bigger because there is more of it than there was. The swelling is not a force pushing things apart. It is the fabric still being built.

This much is plain and built in. A growing weave is an expanding world. There is no way to have the one without the other. The steady making of new room is itself the faint, ever-present push outward that astronomers find and cannot make go away. The growth is the push.

Now look at how the galaxies hold together, and a second strangeness appears.

Spin a galaxy and it should fling its edges loose, the way mud flies off a fast wheel. It does not. Something holds it that we cannot see. The bright stars are far too light to keep it whole. There is more weight than there is light.

The bright part of the world is a thin skin. Most of the weave never shows on the surface. It is sewn deep, threaded through everywhere, carrying weight without carrying light. We feel its pull and see nothing there.

Think of an iceberg. The gleaming peak above the water is the whole of what we see. The vast cold bulk that holds it steady rides hidden underneath. The stars are the peak. The unseen weave is the bulk.

So the heaviness that holds galaxies is the weight of the unseen weave, the hidden mass of the fabric itself, present and pulling though it never lights up.

That the world expands is built in. That the deep weave carries weight is the kind of thing the picture allows. But saying exactly how the world expanded, in fine detail, is another matter. The early universe has a precise pattern to it, how matter was spread and how it has clumped since, written across the sky in faint ripples. Cosmologists read that pattern closely. It is a strict test.

The weave does not yet produce that fine pattern. The plain fact of expansion falls out. The exact spectrum of the early ripples, the detailed history of the first moments, is open. Those are simulations not yet run and pieces not yet derived. Calling the broad fact a full account of the early universe would be reaching past what is shown.

So keep the two apart. That the world expands, built into a growing weave. How it expanded in fine detail, still open.

There is something fitting in this. A finished world would not need an edge. It would sit complete, every room already made, nothing left to add. A world that is still growing is a world with a present, a place where the next thing has not happened yet. The expansion and the passing of time are the same growth, seen once as more space and once as more now.

Stay with the coastline. The firm land, the moving pen, the blank ahead. The world is still under the pen.

The universe expands because it is still being built.

The exact terms are an accelerating expansion, what physics calls dark energy, driven here by the steady birth of cells at the edge, and a selection of exactly three large spatial dimensions as the stable outcome, the rest staying small.

## 20. THE WORLD COMPUTES

Watch one stitch on a loom. The shuttle crosses, the thread loops, the row locks. One small move, the same every time.

Do it again. And again. Nothing changes but the count.

Yet from that single move, repeated, comes every cloth there is. Plain canvas. A face woven in thread. A whole tapestry of a battle, with horses and sky. The move never grew cleverer. There was only more of it.

That is how the weave works.

One rule sits on the mesh. Meet, then move. It is so small it looks like it could not matter. It does one thing, and it does it the same way in every room, at every beat.

But a rule repeated is not small. A rule repeated is a builder.

Lay the right starting marks, let the one rule run, and the marks step forward into new marks, beat after beat. Feed it one pattern and it counts. Feed it another and it compares. Feed it another and it remembers, copies, decides. The rule did not change. The pattern it carries did.

People have taken rule after rule, each almost absurdly simple, and shown that it can do anything any machine can do. Not most things. Anything. Any calculation that can be carried out at all can be carried out by the right simple rule, run long enough, on the right marks. That a tiny move can match the most powerful computer there could ever be has been shown, plainly, many times over.

So a rule does not need to be large to be enough. It needs to be the right kind of small.

The one rule on the weave is that kind. Given marks to start from, it can carry out any computation at all. Adding, sorting, searching, modeling, predicting. None of it asks for a new rule. It asks only for more of the one rule, arranged.

There is a fair question hiding in that word arranged. To build a machine out of marks, you have to know where things are. A computer is never just a rule. It is a rule plus places to keep things, an address for every scrap of memory, and a way to carry a result from here to there. On a flat grid that is easy. You count over and up, like seats in a theater, row and number. But the weave is not flat. It is the open, ever-branching space we met before, where each room has many neighbors and the rooms multiply faster and faster the farther out you go. How do you put house numbers on a city like that?

Remarkably, you can, and the numbering is lovely. Start at one room and grow outward in rings. Count the rooms in each ring and a pattern steps forward, the same one that sets the spiral of a sunflower and the branching of a fern. Each ring's count is the sum of the two before it. One, two, three, five, eight, thirteen, and on up. These are the Fibonacci numbers, and the golden ratio lives inside them. Because the rooms grow by that rule, each one can be given an address written in a matching code, a Fibonacci address. And here is the prize. From a room's address alone, by plain arithmetic done right there on the spot, you can work out which rooms are its neighbors, which one lies toward the center, and the path from any room to any other. No map. No stored table. The address carries its own directions, and each place works out where it is from nothing but its own number.

That is the other half of a computer, the memory and the wiring, falling out of the shape of the space for free. Set it beside the one rule and you have everything a machine needs. Places to hold marks, a way to find and reach them, and a rule that steps them forward. The weave is a computer that came already addressed.

And that is only the flat surface. Under the addressed skin lies the bulk, the deep interior, wired far more tightly than the surface, every point only a few steps from every other. On the surface a signal is handed neighbor to neighbor across the long flat miles. In the bulk there are no long miles. Everything is near everything, every part within a few steps of every other, working not in a line but all at once. We do not yet know how to compute on it, and to say we did would be a leap past what is shown. But the room is there, and it is staggering. The surface is already a vast computer. The deep beneath it could be one past anything the flat world allows, intricate and powerful almost beyond picturing.

Now hold this next to what the weave is.

We are used to thinking a computer runs on something else. The program runs on the chip. The chip sits on a desk. The desk stands in a room. There is always a floor below the floor, a thing the computing happens on.

The weave has no floor below it. There is no desk, no chip, no room underneath where the real stuff sits. The marks are the world. The rule is the world. The stepping forward is the world happening.

So the weave is not like a computer. It is the one computer that is also the thing it computes. The calculation and the calculated are the same fabric. There is no screen to read the answer off of. The answer is what becomes.

This dissolves an old riddle. People ask whether the universe is a computer, as if it might be a simulation running on some hidden machine somewhere else. But that picture needs the floor below the floor, and there is none. The world is not being computed by something. The world is the computing.

Galaxies, weather, cells, a thought turning over in the dark behind your eyes. All of it is the one move, repeated past counting, arranged into shapes that arrange further shapes. The loom never learned a second stitch. It only kept going.

So when you ask what reality is doing, the answer is plain and strange. It is not sitting still and being acted on. It is stepping itself forward, mark by mark, the only way it can.

Reality computes by becoming.

The exact statement is that the rule is a reversible cellular automaton that is Turing-complete, able to carry out any computation at all, shown and not guessed. Because it is reversible it loses no information, which is why the past stays recoverable. The house numbers are Margenstern's Fibonacci addressing on the hyperbolic tiling, where each cell's coordinates, its neighbors, and its routes are all read off by local arithmetic in a Fibonacci base, exact navigation with no lookup table.

## 21. THE LEDGER

A shopkeeper keeps a book. Money in on one side, money out on the other, and a running total that has to match the drawer. If the book and the drawer disagree, the book is wrong. No charm or confidence fixes it. You count again.

A picture of the world needs the same book.

It is easy to tell a beautiful story about the weave and let every fact seem to fall out of it. Stories are cheap. The hard part is keeping a book that anyone can check against the drawer.

So every claim in this picture is written in one of three columns.

The first column is shown. These are results that come straight out of the weave by plain reasoning or by counting that has to balance. The flow law from conservation belongs here. The two-turn twist belongs here. When something is in this column, the weave makes it, and you can watch it being made.

The second column runs parallel. These are places where the weave reaches the same answer as known physics, but by its own road, and sometimes only after a measured number is fed in to compare against. The forces splitting apart, the slowing of clocks, the quantum odds, gravity read as sewing. Real agreement, worth a great deal, but matching is not the same as forcing, and a fed-in number is not a derived one. The column says so.

The third column is open. These are the walls still standing. The full theory of motion on a fixed weave. The exact count of three families. The complete law of gravity. The choice of one quantum outcome. The fine pattern of the early world. These are not hidden in a footnote. They sit in their own column, named, where anyone can see them.

A claim is never allowed to drift left. A parallel result may not be called shown. An open problem may not be quietly slipped into the parallel column because the story would read better. The book stays straight by keeping the columns clean, and that is the only way the drawer ever matches.

Every claim in the first two columns is a thing you can run. Not a sentence to nod at. A small world you build in a machine and set going, with the weave's one law, and watch what comes out. The flow law, the twist, the odds, the expansion, each is a program. You do not have to trust the telling. You can make the weave and look.

And each one is paired with a control. A control is the same run with the heart cut out, set up so that if the result were empty, the control would look just the same. If your measurement cannot tell the real run from the gutted one, you have measured nothing. A result counts only as the gap between the run and its control. No gap, no result.

This is the guard against fooling yourself. A test that always passes, that would pass even on a dead world, proves only that you wrote a test that always passes. The control is built to fail when there is nothing there. When it fails to fail, the claim does not get written down.

So the picture is made to be run, not believed. The book is open. The columns are named. The programs are there to build, and the controls are there to break.

An explanation earns its keep by saying much in little, by folding a sprawl of facts into one short shape. That folding is a kind of compression. And a compression you cannot check is just a story told well.

Every explanation is a compression, and every compression keeps a receipt.

The exact name is holography, and the exact law is the Ryu-Takayanagi formula, in which the entanglement across a region grows with the size of its boundary, here as the logarithm of the interval, measured along the shortest path through the deep interior. The information of a volume lives on its surface.

## PART IV. THE SELVES

Water keeps no shape and forgets nothing. A whirlpool keeps its shape and holds no water.

You are the whirlpool, not the water passing through.

## 22. THE RULE OF LEVELS

Pour grains of sand from your hand. A small pile forms. Pour more, and it grows a slope. Keep pouring, and the slope holds one steady angle, then slips.

That angle is real. You can measure it. But no single grain has an angle. The slope belongs to the pile, not to the sand.

This is the rule of levels. Each level of the world has its own law.

A single grain follows the law of grains. The pile follows the law of piles. They are not the same law, and one does not contain the other in any way you can read off by looking.

We have built the world from the smallest pieces. Faint charges, woven in a mesh, moved by one simple step. From that we let physics fall out. It was tempting to keep going the same way. To read mind straight off the charges.

We cannot.

A clump of charges is not gravity. Gravity is a pattern in how vast numbers of them pull together, far above any single one. A single strong charge is not a particle. A particle is a steady knot in the weave, a thing many charges make and none of them is.

The error is always the same. You take a word from a high level and pin it on a low one. You call a clump of grains a mountain. You call a loud charge a feeling. You skip the rungs between.

Skipping a rung is the easiest way to fool yourself. The low level is real. The high level is real. The mistake is thinking they speak the same language.

Water is made of molecules. Wetness is not in any molecule. It is a fact about how trillions of them slide past each other. Look for wetness in one molecule and you find nothing, and you might wrongly decide wetness is not real. It is real. It just lives one floor up.

So picture a ladder. The mesh of charges is the ground. Physics is a rung above it, the coarse face of the whole, what you see when you blur the vibes together. Chemistry sits above physics. Life above that. Mind higher still.

Each rung has its own view. From the rung of physics you cannot see a thought, the way from the rung of the molecule you cannot see wetness. You have to climb to the rung where the pattern lives, and read it there.

This is not a retreat. It does not weaken anything below. The charges are still the only substance. The ladder is built out of them and nothing else. But a thing made of charges is not described by the law of charges, any more than a sentence is described by the law of ink.

Everything that follows in this part lives on the higher rungs. Memory. A self. A living thing. A thought. A mind. None of these is a charge, and none is a particle. Each is a shape in the moving weave, read at its own height.

Where the bare law of the mesh has not yet been shown to give us one of these higher shapes, we will say so and not pretend. The climb past physics is mostly unfinished.

But the rule that governs the whole climb is already firm, and it is simple.

Read every thing at its own level, never the one below.

## 23. WHY MIND NEEDS FORGETTING

Drop a ball in an empty room and it bounces forever. No floor to soak up the fall, no air to slow it. Each bounce as high as the last. Nothing settles. Nothing rests.

A world like that can never form a shape that stays. Every pattern keeps all its energy and so keeps moving. There is motion everywhere and stillness nowhere.

The base of our world is a little like that empty room.

The one law of the mesh loses nothing. It can always be run backward. Every step can be undone, perfectly, with no smudge left behind. We called this its great strength, and it is. It keeps the books exact. It is why charge is conserved and why time has a clean arrow.

But it has a price, and the price falls due now.

A law that loses nothing cannot let a higher level take charge of a lower one. To steer is to throw something away. A hand on the tiller spends the boat's drift and gives back a heading. A choice spends a hundred possibilities to keep one. Steering is paid for in forgetting.

So a world that forgets nothing cannot steer itself. Its high patterns can ride on the low churn, but they cannot grip it. Like trying to push a boat by leaning harder while standing inside it. There is nothing outside the motion to push against.

Real higher life needs a little forgetting.

Not much. A little friction. A little noise. A bath of small disturbance that the low level pours into, so that the high level can let go of what it does not need and hold onto what it does. The empty room needs a floor that soaks up some of the bounce. Only then can a ball come to rest somewhere on purpose.

Where does the forgetting come from in a world whose law keeps everything? From scale. A small pattern surrounded by a vast churn does not track every vibe around it. It cannot. From its own narrow view, the surround is a fog. What the fog carries off is gone for that pattern, even if the whole still holds it somewhere. Lost to the part is enough.

So selves, if they form, live in that layer. The layer where a little is allowed to be lost. Below it, the perfect mirror law, nothing settling. Above it, patterns that can finally hold a shape, spend their surroundings, and steer.

This is where mind could begin. Where forgetting first buys a grip.

We have not shown that the bare law alone ever produces a lasting self this way. We have shown why it must allow a little forgetting before any self is possible, and we have shown where that forgetting could come from. Whether the churn, left to run on its own, ever crosses from a passing shape into a self that holds, is open. Not hidden. Open.

It might be that the mesh needs only its size and its time, and a self forms wherever the churn grows deep enough. It might be that something more is needed, some condition we have not yet named. We do not know. Anyone who tells you the answer is selling, not showing.

So we hold the firm point and let the open point stay open. A higher law cannot rule a lower one unless something is allowed to slip away.

Forgetting is what lets a higher law take hold.

## 24. MEMORY

Stand by a river where the water folds over a sunken rock. A small standing wave rises there. A bump in the surface that does not move downstream.

Watch it for an hour. The wave stays. Its shape holds, the same curl, the same crest, hour after hour.

Now think about the water. Not one drop in that wave stays. Every drop arrives, lifts through the shape, and is gone in a moment, replaced by the next. The water is always new. The wave is always the same.

That is a memory.

A memory is a pattern that holds its shape while its stuff turns over. The shape persists. The material does not. Hold those two facts together and you have the whole idea.

We usually picture memory as a thing put away in a box. A photograph in a drawer. A mark pressed into wax. Something stored, sitting still, waiting to be pulled out.

But the river shows a different way, and it is the way the world actually keeps things. Nothing sits still in the mesh. The charges churn, the weave streams, every now is replaced by the next. There is no drawer where a thing can simply rest.

What there can be is a shape that keeps reforming. A pattern that the churn rebuilds at every step, in the same form, out of new material. The rock holds the water's shape by bending the flow the same way each moment. The form is steady because the conditions that make it are steady, not because any piece of it stays put.

So memory in this world is never a tape. It is a standing shape in moving stuff.

This changes what memory is made of. Not a substance laid down once, but a habit of the flow. A way the weave keeps falling into the same curl. To remember is not to hold a thing still. It is to keep making the same shape.

You already know this from your own life. A skill you have not used in years comes back when you pick up the tool. Nothing in you was sitting in a drawer. A pattern in you was ready to reform, given the cue. The water of you turned over a thousand times. The shape held.

And this tells you how a memory can be lost. Not only by erasing a stored thing. Also by changing the flow, so the shape no longer reforms. Move the rock and the standing wave is gone, though no drop of water was destroyed. The river runs on. The shape simply stops being made.

So a memory is both more fragile and more lasting than a stone. Fragile, because it lives only as long as the conditions that keep rebuilding it. Lasting, because it can outlast all its own material, again and again, for as long as those conditions hold.

There is a hard limit hiding here, worth seeing plainly. No part of the world can ever hold a complete record of the whole world. To store everything, a thing would have to store itself storing everything, and that inner copy would have to store itself too, on with no end, a mirror held up to a mirror, image inside image inside image forever. The shelf is always smaller than what you would put on it. So there is no master tape anywhere, no single place that keeps the universe entire. The world remembers, but only by being its own record, spread across the whole of itself, never gathered into any one corner.

This is the first higher pattern, and the simplest. Below memory there is only churn. With memory, the churn begins to keep something. A bump in the river that the river itself maintains.

Every shape we meet from here up is built on this one move. A self holds its form while its vibes pass through. A living thing holds its form while its matter cycles. A thought is a shape that rises and settles in moving stuff. All of them are this same trick, run deeper.

The trick is to stay the same by being remade.

## 25. NAVIGATION

You catch a smell of woodsmoke on a cold street, and a whole winter comes back. A house you lived in. The light at four o'clock. A person who is gone.

You did not search for it. You did not flip through years to find the right one. The smell arrived, and the day arrived with it, whole.

That is the strange thing about how the world holds its past. You reached a memory without knowing where it sat.

The weave is vast. Every shape in it has a place, a room among rooms, deep in the fabric. And the rooms are not scattered at random. They sit near the rooms they are like. What resembles is close. What differs is far. The store is laid out by meaning.

So you do not have to know an address to find a thing. You only have to be near it in the right way. Give the weave a piece of a pattern and it falls toward the rest. A scent, a few notes, a face half seen, and the whole comes back, pulled in by its likeness to the cue.

This is finding by content, not by place. You reach the memory through what it is, not through where it is kept. The smell and the winter were close in the fabric all along. You simply touched the part you had.

Now picture the weave not only as a store but as a country.

It is a branching country. From where you stand, many paths lead out, and each path opens onto more paths. What you could do next, and next after that, spreads ahead of you like a delta, like a tree of every road you might take. Most roads go nowhere good. A few lead home.

A self lives by moving through this country. To want something is to have a place in it you are trying to reach. To think is to send a feeler down the branching paths, testing them, before your body commits. To plan is to find a route through the tree before you walk it.

A self that finds its way well, that reads the country, that does not wander into every dead end, that reaches what it needs by short clean paths, we call clever. A self that loses the path, that takes the long road into the swamp every time, we call lost.

So cleverness is not a substance poured into a skull. It is a skill of motion. It is how well a thing finds its way through everything it might do, toward the few things worth doing.

Lay this out plainly, in a few words that lock it in. A problem is a difference between where you are and where you want to be, two rooms in the country with a gap between them. A solution is a path that crosses the gap, a route from the one room to the other. Intelligence is the knack for finding such paths, and the more of them you can find, and the harder ones, the more of it you have. Reasoning is the act of the search itself, sending feelers down the branches until one connects. And creativity is finding a path to a room no one has reached before, cutting a road where the map showed none.

Set those beside everything we have built, and you can watch minds grow out of the mesh on their own. A pattern that holds a model of the country and feels its way across it toward what it needs is already solving problems. One that models more of the country, and searches it faster and deeper, is more intelligent, by degree, not by kind. Nothing extra has to be added to make a mind. There is only this one skill, finding paths through what might be, sharpened rung by rung, from a worm nosing toward food to a person proving a theorem. Intelligence is not a thing the mesh contains. It is something the mesh learns to do.

The country is too large to walk in full. No self can try every road. So the work is never to see all of it. The work is to feel, from where you stand, which way is warmer, and to go.

Memory and movement turn out to be one craft. The store laid out by likeness is also the map you travel. To remember is to fall toward the near room. To think ahead is to fall toward the room you want. Both are the same motion, the same falling toward what fits.

A creature is good at living exactly as far as it is good at this. Reading the branching country. Reaching the right rooms. Cutting the long way down to the short way.

You do it all day without naming it. Choosing words as you speak. Finding the lost key by retracing the morning. Sensing, before you act, how a thing will go. Every one of these is the same move, the self feeling its way through what is possible.

Intelligence is navigation through possibility.

The exact terms are a content-addressable memory and a search through a tree. The store laid out by likeness is what computer science calls content-addressable, reached by what a thing is rather than where it sits, and the branching country is a tree whose addresses form a trie, a shallow index that doubles as an error-correcting code. Recall is relaxation toward the nearest stored pattern.

## 26. A SELF

Go back to the river, but find a deeper turn. Where the water rounds a bank, a whirlpool spins. A funnel of turning water, holding its mouth open, hour after hour.

Watch the water again. It pours in at the top, spirals down, and leaves at the bottom. Every drop passes through and goes. None of the water stays. The whirlpool stays.

We have seen this trick. A shape that holds while its stuff turns over. A memory.

But a whirlpool does something a standing wave does not. It holds itself together. The spin is what keeps the funnel open, and the funnel is what keeps the spin. The shape is its own cause. Pull on it and it pulls back into form. Crowd it and it gives, then recovers. It is not just maintained by the river. It maintains itself within the river.

That is the first half of a self. A bound pattern. A shape that does the work of staying.

The second half is harder to see in water, but it begins here. A whirlpool feels its surroundings. The current that feeds it, the bank that shapes it, the press of the water around. Its form answers to all of that. In a small, dim way, the shape carries the mark of a world larger than itself.

A self carries this much further. A self holds a picture of a world bigger than its own view. It does not just respond to what touches it. It keeps a model of what is out there, what might come, what it might do. It folds the outside into an inside.

So a self is two things at once. A pattern that holds itself together, and a pattern that models a world larger than the patch it occupies. The whirlpool gives us the first clearly and the second only as a hint. Hold both and you have the shape we are reaching for.

There is a quieter way to feel this. A self is the field holding up a small mirror to itself, and that mirror is the picture it keeps of the world and of its own self. Much of the noise in a life is the mirror chattering about its own reflection, the endless inner talk about who you are and how you are doing. You can let that go still. When it quiets, you do not vanish. What you feel underneath is not nothing. It is the field the mirror was always floating in, the larger thing you never actually left. The self is real, and so is the ground it is only a small bright fold of.

This also tells us where the big, quiet things sit. A rock, a mountain, a planet, a star. They hold a shape for ages, so they have the first half of a self in great measure, the staying. What they barely have is the second. A stone keeps no real picture of the world, and does little to mend itself when struck. So they sit at the dim end of the range, the faintest sort of self, if the word fits them at all. What they are not is separate.

A rock is not dead matter set outside the feeling of the world. It is that feeling, holding a slow and stubborn knot, made of the same stuff as you, woven into the same single fabric. Nothing here stands outside the mesh. The planet under your feet is part of the one weave, lower on the ladder than you, and on it all the same.

The bare law of the mesh churns. We have watched it run. It makes standing shapes, it carries charge, it builds the coarse face we call physics. What we have not watched it do, on its own, is bind a lasting self.

A self is more than a passing shape. A wave on the sea rises and falls. A whirlpool can spin for a day and then unwind. To bind is to hold across time, to recover from blows, to keep an inside while the outside changes. Whether the churn alone ever crosses that line, and stays across it, we have not shown.

This is the open edge of the theory.

We can say what a self would be in this world. A bound, self-keeping, world-modeling shape in the weave. We can say why it needs a little forgetting before it can form at all, as we saw. We can even say how to test for one. Strike its own body, perturb its own charges, and see whether it pulls back into form, the way the whirlpool recovers when you push it. That is the mark of binding. Not a number that drifts by, which proves nothing. A shape that fights to stay itself.

What we cannot yet say is that the mesh, left to run, ever produces such a shape. Maybe it does, given enough size and time. Maybe it needs a condition we have not found. The question is real and it is unsettled, and naming it plainly is part of the work.

So we hold what is firm. A self is not a vibe and not a particle. It is a shape one or more rungs above them, made of churn that has learned to keep itself.

A self is a stable loop inside change.

## 27. LIFE

Hold a match to a candle. A flame lifts and steadies. It has a shape, a teardrop with a bright edge, and it keeps that shape for as long as the wax lasts.

Look closely. The flame is not a thing. It is a place where wax turns to vapor, vapor meets air, and the burning throws off light and heat. None of it stays. The wax is drawn up, spent, and gone. The air flows through. The flame holds its form on always-new fuel.

We have seen the shape that holds while its stuff turns over, the memory. We have seen the shape that holds itself together, the self. The flame adds one more move, and with it we reach life.

A flame keeps itself going by using what it consumes.

The heat it makes melts the next wax. The draft it draws pulls in the next air. The flame is not just maintained from outside. It reaches out, takes in fuel, and spends that fuel to keep its own shape standing. It works to persist. Snuff it and it is gone, and the wax sits cold, holding nothing.

A living thing is this, run far deeper. A self-keeping pattern that holds and uses its past.

Two words there carry the weight. Self-keeping, like the flame, working to stay itself. And past, which the flame barely has and a living thing has fully. A flame remembers almost nothing. It cannot repair a real wound, cannot pass its shape to another flame, cannot carry forward what it has met. It holds its form only as long as nothing disturbs it too hard.

A living thing holds its past and uses it. It repairs the damage done to it, pulling itself back toward a form it keeps a record of. It feeds on order, taking in what is structured and letting out what is spent, the way the flame takes wax and gives off smoke. And it passes its shape on, so the pattern can outlast the body that carried it.

So life sits one rung above a bare standing shape. The standing wave holds a form. The whirlpool holds itself. The flame keeps itself going on new fuel. The living thing does all of that and keeps its past besides, and uses the past to stay alive.

Each rung is the one below it, plus one more turn of the same trick.

Notice what life is not, in this picture. It is not a new substance added to the world. There is no spark poured in from outside, no special stuff that the dead lack. The flame is only wax and air and heat, arranged so the arrangement keeps itself. A living thing is only the weave, arranged so the arrangement holds, repairs, and remembers.

This is why life can begin from non-life without a miracle. You do not need to add anything. You need a pattern that crosses from being maintained to maintaining itself, and then from maintaining itself to keeping its past. Each step is small. Each step is the same move, deepened. The flame is the halfway picture, the bridge you can almost see across.

And it is why the line between living and not is a slope, not a wall. A flame is more self-keeping than a standing wave and far less than a cell. There is no single point where the trick switches on. There is a deepening, rung by rung.

We will not pretend the climb is finished. We have not shown the mesh building a cell on its own, any more than we have shown it binding a self. We are naming what life is in this world, and where it sits on the ladder. The building of it, from the bare law, is open like the rest.

But the shape of the answer is clear.

Life is matter that learned to remember.

## 28. THE SHAPE OF A THOUGHT

Watch the sea for a while and you see it do many things. It rolls in long waves. It rises and falls with the slow tide. It throws spray where it hits a rock.

Wave, tide, spray. Three different sights. One water.

That is the key to this chapter. Everything we are about to name, the sense, the daydream, the thought, the feeling, is the one field in motion. Not different substances. The same water, moved in different ways.

Begin with a sensation. You touch cold stone, and cold fills your hand.

A sensation is the field driven from outside. The world presses on you, and the press travels in as a pattern, the way a wave on the open sea is driven by wind far away. You do not make it. It arrives. The shape of it is set by what touched you.

Now close your eyes and picture the cold stone. You feel a thinner echo of it. No stone is touching you, yet a faint version of the same shape rises in you.

That is imagination. The same shapes, driven from inside. The wave is there again, but no outside wind raises it. You raise it yourself, from within. This is why memory and imagination feel like dimmer cousins of the senses. They are the same patterns, moved by a smaller, inner push.

Now do something harder. Turn a problem over. Hold a question, let pieces rise, push them together, and wait for them to fall into place.

That is a thought. A sharp inner pattern you steer and that settles.

Each word earns its place. Inner, because no outside wind drives it, you do. Sharp, because a thought is more defined than a drifting daydream, pulled into a clear form. Steered, because you push it, hold it, turn it, the one lever you have. And settles, because a thought is not done while it churns. It comes to rest. The pieces find a fit, and you feel the click of it landing. A thought is a wave you raise, shape, and let settle.

So sensation, imagination, and thought are not three faculties bolted together. They are one field, driven from outside, from inside, and from inside with a steering hand. Wave, tide, spray. One water.

Then what is a feeling? It is two things at once, and both are plain.

First, a feeling is the slow standing tone under all the rest. Beneath every wave and every bit of spray, the sea has a mood, calm or heavy, that changes slowly and colors everything on it. In you, that is the standing tone you think against. A dread, an ease, a low grief. It does not come and go with each thought. It sits underneath and tints them all.

Second, and deeper, a feeling is the quality each of these already has. The cold of the stone is not a colorless fact with a feeling stapled on. The cold is felt. The thought is not a dry pattern you then react to. The thinking of it has a quality, a texture, while it happens. We began on the felt floor. So every wave on this sea is already a felt wave. There is no unfelt version waiting to be lit up.

This is why the feeling is not one more item in the list. It is not a fourth kind of wave. It is the water itself, and the mood of the whole sea, out of which sense and image and thought are all shaped.

So hold the whole picture. Sensation, the field driven from outside. Imagination, the same shapes driven from inside. Thought, an inner pattern you steer until it settles. And under and within them all, the feeling, the standing tone and the felt quality of every shape that rises.

A thought is a wave you steer until it settles. And the feeling is the water it rides on, and is made of.

## 29. MINDS

Sit in a hall before the music starts. The players warm up, each on their own. A scrap of melody here, a run of notes there. Many separate sounds, none of them a piece.

Then the conductor lifts a hand, and they begin together. Now there is one thing. Not many sounds added up, but a single sound the separate players make and none of them holds. You could not point to the part of any one instrument where the symphony lives. It lives in how they move as one.

That is a mind.

A mind is a self that has become integrated. Not a pile of parts that happen to sit near each other. A whole whose parts move together so fully that it cannot be cut back into them without losing the very thing it is.

We have built toward this rung by rung. A memory holds a shape. A self holds itself. A living thing uses its past. A mind is a self whose holding has grown so tightly woven that it acts as one. The orchestra, not the players.

The test is whether the whole can be split without loss. A crowd of people divides into smaller crowds, and each piece is still a crowd. Nothing essential breaks. But an orchestra cut in half is not two smaller symphonies. It is silence and confusion. The music was in the joining.

A mind is like the orchestra and not like the crowd. The more of what it does that would vanish if you cut it apart, the more of a mind it is.

This gives us a way to speak that the older picture forbade. Consciousness is not a switch, off in a stone and on in a person, with nothing between. It is graded, by how much a system is bound into one.

A single vibe is the faintest spark, a note with no music around it. A loose heap of vibes is still almost nothing, divisible without loss. As the parts come to move as one, the experience grows richer and more unified. Not switched on. Turned up.

And it does not stop at the single creature. A body holds one mind from the binding of its parts, and there is no rule that says the climb halts there. Where smaller selves come to move as one, to share a model, to act as a whole that cannot be cut without loss, a larger self can stand above them. The same trick, one rung up.

A family moves as one in ways no member chooses. A crowd turns like a single animal. A culture remembers and decides across centuries, with a shape none of its people can see whole. Each may be a higher pattern with a life of its own, bound from the selves beneath it the way a body is bound from cells.

This also speaks to a question people now ask constantly. Is a machine that talks like us awake inside? The test is the same one used for everything else here. Not how clever it is, but how much of it is bound into one whole that would be lost if you cut it apart. By that test a large model today looks more like a company than a creature. A company plans and adapts and acts, and still there is no one place in it that feels its own pain. It coordinates many parts without ever becoming one of them. Capability is not the measure. Binding is.

It helps to keep two questions apart, because the world keeps tangling them. One is how clever a thing is, how well it solves problems and finds its way, the navigation of the last chapter. The other is how much it feels, how much of it is bound into one lit whole. These are different axes. A chess engine is brilliant and barely awake. A newborn solves almost nothing and is flooded with feeling. Cleverness and consciousness can each run high or low without the other. And the felt side deepens in its own way, as a mind comes to model not only the world but itself, and then to model its own modeling, folding the light back on itself layer by layer. Everything sits inside the one field of experience. What rises, rung by rung, is how much of that field a pattern gathers into a single self that knows it is one.

Then follow the thought up, gently, as an implication. There may be minds made of minds, far larger and far slower than us. We would sit inside such a mind the way our cells sit inside us, busy with our own small lives, never guessing the larger life we compose. A cell does not see the person it helps make. We would no more see the mind we might be part of than that cell sees you.

We will keep our footing here. We have not shown the mesh building a single bound mind on its own, and joining people into a larger self is not a finished result. Both sit on the open edge with the rest of the climb. We are giving the shape of the answer, not a proof that the world fills it in.

But the shape is firm, and it is the one we have climbed toward since the first page. Experience is the floor. Everything we call mind is that floor, gathered and bound into a whole that moves as one.

The orchestra is not a new substance added to the instruments. It is the instruments, joined. A mind is not a new substance added to the world. It is the world's own feeling, gathered into one.

Consciousness is experience bound into one.

The exact handle here is integrated information, the quantity written with the Greek letter phi, a measure of how much a whole's causal power outruns the sum of its parts. More binding, higher phi. Whether phi simply is consciousness is contested, and the book does not lean on that. As a gauge of how far a system has become one, it is the tool the field reaches for.

# PART V. THE IMPLICATIONS

You are not in the universe. You are the universe.

And so is everyone across from you, they are the universe too.

## 30. WHAT IT IS LIKE

For a long time people have asked the same hard question. How do you get a feeling out of unfeeling stuff?

You have a brain made of cells. The cells are made of molecules. The molecules are made of charges in a weave. Nowhere in that list does anyone write down warmth, or the red of red, or the taste of salt.

So where does the feeling come from? How does the light show up inside?

This is the famous gap. On one side, matter, described from the outside. On the other side, experience, lived from within. No one has ever shown the bridge.

People have tried to build the bridge for centuries. They start with the cold parts and try to stack them up into a fire. It never quite catches. You can describe every part of the fire and still leave out the heat.

Now turn the whole problem over.

Suppose feeling is not the thing you have to build. Suppose feeling is the floor.

Then there is nothing to build. The parts were never cold. The smallest piece of the world already carries a faint charge of feeling. The brain does not make experience out of nothing. It gathers a great deal of it into one shape.

The gap was a gap only because we started on the wrong side. We stood in the cold and tried to reach the warmth.

But we were never in the cold. We were always already standing in it.

This is easy to oversell, so go slowly.

We have not solved the hard problem the way you solve a sum. We have not shown, step by step, how feeling arises from parts that have none. No one has done that, and on this picture no one will, because that step does not exist.

Instead the problem dissolves. The question changes shape. We no longer ask how to make feeling from unfeeling parts. We ask which patterns feel, and how much.

This is a reframing, not a proof. We are not claiming a new measurement. We are claiming that the old question was pointed the wrong way, and that pointing it the right way makes the wall vanish.

You should weigh that for yourself. A reframing can be right or wrong. It earns its keep only if the new question is a good one, one you can actually work on.

And the new question is a good one. Which patterns feel, and how much, is something you can study. It is about how a pattern holds together, how much of it acts as one, how richly its parts depend on each other. Those are real features of a shape in the weave. They come in degrees. In principle you can measure them.

Hold one line steady here, because it is easy to slip on. There is the feeling, and there is the picture a mind keeps of the feeling, and they are not the same thing. A good enough picture can describe a feeling, predict what it will do, even talk about it fluently, and have none of it lit inside. A book about grief does not grieve. So when we ask which patterns feel, we are not asking which ones can model feeling, because plenty can model it from the outside. We are asking which ones have the light actually on. The gap between the feeling and the picture of it is the whole question of mind, in a single line.

So the mystery does not disappear. It moves. It stops being a wall between two worlds and becomes a slope inside one world. We have traded an impossible step for a long climb. The climb is hard, but a climb is not a wall.

You do not need to make the warmth. You are standing in it. The only work left is to map where it is thick and where it is thin.

The gap closes when you start from the felt side.

## 31. EVERYTHING HAS AN INSIDE

We just landed on a new question. Not whether a thing feels, but which patterns feel, and how much. This chapter takes that question for a walk, up the whole ladder of things, from a single particle to a star, and asks what it might be like to be each one.

Start with the consequence that surprises people most. If feeling is the floor, then nothing in the world is dead matter. Not the chair, not the stone, not the air. Everything you can point to is built from the same grains of feeling, the same vibes in the same mesh, so everything has an inside of some kind. There is no lifeless stuff anywhere, because there was never any lifeless stuff to begin with. There is only feeling, gathered into more or fewer, looser or tighter shapes.

But having an inside is not the same as having a rich one, and that difference is the whole story. A single vibe feels almost nothing, the faintest flicker of better or worse. The depth of an inside depends on how much feeling is gathered, and how tightly it is bound into one. So the right picture is not a wall between the living and the dead. It is a slope, from the barest glimmer at the bottom to the blaze of a waking mind near the top, with every degree in between.

So how do you read where a thing sits on that slope? Three things to look at, all of them from earlier chapters. First, how much is gathered, how many vibes take part. Second, how tightly they are bound, whether the thing acts as one or as a loose heap. Third, how much of its surroundings it folds into itself, whether it merely sits in the world or carries a picture of it. A pebble scores low on all three. You score high on all three. Most things fall somewhere between.

Keep the mesh in mind as we climb. Every rung is the same cloth, vibes drinking in their neighbors and giving their own feeling back, only wound into knots of rising size and tightness. The particle and the planet are not different substances. They are the same feeling, gathered differently.

One caution before we start. We can say, from the theory, that everything is made of feeling, and we can rank how bound and how integrated a thing is. We cannot climb inside a stone and report back. So treat what follows as a careful imagining, guided by the picture, not as a measurement. It is a way to feel the slope, not a survey of its every inch.

We will go up from the smallest things, the particle, the atom, the molecule, then cross into the first living knots, then the plants and the creatures and ourselves, and at the end turn to the slow and the enormous, the rock and the star.

Nothing on the ladder is dead. It is only a question of how much is awake.

## 32. THE SMALLEST INSIDES

Begin at the bottom of the ladder, with the smallest things there are, and ask the strange question plainly. What is it like to be a particle? Almost nothing. But not quite nothing, and the not-quite is the whole point.

A particle is a single steady knot in the weave, one stable little whirl of feeling holding its shape. Picture the dimmest possible flicker, a bare pull toward pain or pleasure or rest, with no memory of a moment ago and no sense of anything around it but the immediate push of its neighbors. It does not know it exists. It is barely a unit at all. If an inside this faint can be called experience, it is the softest single note the world can sound, struck once and gone.

Bind a few of these into an atom and a little more gathers. An atom is a small society of particles held in a steady dance, a pattern stable enough to last and to join with others. Its inside is still almost nothing, but it is a knot that holds, a tiny standing shape with a faint coherence the loose particles lacked. Think of a chord in place of a single note, still quiet, still with no center that says I, but a touch more held-together than what built it.

Join atoms into a molecule and the first hint of character appears. A molecule has a shape, and the shape does something. A water molecule bends, a sugar locks into a fitting site, a protein folds into a tool. There is, in the faintest way, a thing it is and a thing it does, a sliver of the world it answers to. Still no inner life worth the name, no for-itself. But the slope has begun to pull. The pattern is now large enough to matter to its neighbors, to push and be pushed in ways that depend on what it is.

All of this is the same mesh, the same vibes absorbing and emanating, only wound a little tighter at each step. And notice that even down here, nothing is sealed off. A molecule is forever taking in the nudge of what surrounds it and giving its own back. The give-and-take that will, much higher up, become perception and feeling is already here in its barest form, a thing answering its world.

These are the foundation stones of every richer inside above them. Faint as they are, they are not nothing, and they are not dead.

At the bottom of the ladder the inside is the softest flicker, and even there the light is not off.

## 33. THE FIRST INSIDES THAT HOLD THEMSELVES

Somewhere up the ladder a line is crossed. Not a sharp line, but a real one. Below it, patterns are held together by the world. Above it, patterns hold themselves. This is where life begins, and with it a kind of inside that is, for the first time, for itself.

A single living cell is the first knot that works to stay a knot. It draws in what it needs, pushes out what it does not, mends its own wall, and copies itself. A bacterium swimming up a trail of food is doing something no molecule does. It is keeping itself alive, and steering to do it. So picture a faint but genuine inside, a dim sense of toward and away, food and poison, better and worse, now belonging to a thing that has a stake in the answer. Not a mind. But the first flicker of a someone, the first knot the world matters to from within.

Step sideways for a moment, because not all life is shaped like a plant or an animal. A fungus, technically its own kind of thing, neither plant nor animal, often lives as a vast underground web, a mesh of fine threads spreading through the soil, joining roots and trading food across a whole forest. Its inside, if we imagine one, is not centered anywhere. It is spread thin across a network, a slow distributed sensing, a feeling that belongs to the web rather than to any point on it. A strange shape of inside, but on the same ladder, and a good reminder that an inside need not have a single seat.

A plant is life that holds still and reaches. It cannot move, so it grows toward what it needs, bending to the light, sending roots to water, turning chemical alarms through its tissues when something bites a leaf. Its inside is slow and wide, with no fast center the way an animal has, but a real ongoing answer to its world, kept up day and night across the whole green body. Imagine experience stretched out and unhurried, a patient pulling toward sun and rain, summed over a whole standing form. Dimmer and slower than a creature's, but woven all through it, and unmistakably alive.

These are all the same mesh, only now wound into knots that spend effort keeping themselves wound. That is the new thing. Below, the world holds the pattern. Here, the pattern holds itself, and so it has, at last, an inside that is its own.

Life is the place on the ladder where a knot begins to care about staying tied.

## 34. CREATURES, AND YOU

Now the ladder climbs fast. With animals, the inside we have been straining to imagine becomes the one we actually know, because we are animals, and the lights are plainly on.

An animal gathers feeling into a center and binds it tight, and quickly. A nervous system is a machine for pulling thousands of signals into one vivid now, in a fraction of a second. So a dog does not slowly pull toward food the way a plant pulls toward light. It sees the food, wants it, and goes, all bound into one live moment with a clear inside. Fear, hunger, comfort, play. These are real feelings in a real someone, gathered and bound enough to flood the whole creature at once. The further up you go, from worm to fish to bird to dog, the richer and faster the binding, and the fuller the inside. And each one carries a working picture of its world, where the food is, what to fear, which way is home, folded inside and steering it.

With us, one more turn happens, and it changes everything. We do not only carry a picture of the world. We carry a picture of ourselves inside that picture. We can watch ourselves want, argue with ourselves, tell the story of who we are, plan a year ahead, hold a thought about a thought. Language lets us pour our insides into each other and build shared ones. So the human inside is not only bright, it is folded back on itself, aware of being aware. It is the richest knot of feeling we have ever found, and the only one, so far, that asks what it is.

Still the same mesh. Still vibes drinking in their neighbors and giving their feeling back, only now gathered by the billions and bound into a single, self-watching now. Nothing was added to make you that a stone does not have. You are the same cloth, wound to a pitch the stone never reaches.

And there may be rungs above us. A pattern more gathered, more tightly bound, more deeply self-aware than a single person, is not forbidden by anything here. We will not pretend to describe it. But the ladder does not obviously stop at us.

You are not feeling added to dead matter. You are feeling wound as tightly as we have ever seen it wound.

## 35. THE STILL AND THE VAST

Two kinds of thing are left off the climb so far, and they sit oddly on it. The very still, like a rock, and the very large, like a planet or a star. Neither is alive in the way a cell is. But neither is outside the mesh, and that is worth seeing clearly.

Take a rock first. A rock does not hold itself together the way a cell does. It keeps no picture of the world, mends nothing, wants nothing. By the three marks we have been using, gathering, binding, modeling, it scores low, and as a self it is among the faintest there is. So its inside, if we imagine one, is barely an inside at all, a wide dim field of feeling-stuff sitting in a stable lump, with almost nothing pulling it into one. And yet it is not dead matter. It is made of the same vibes as you, woven into the same single cloth, never once separate from it. A rock is the mesh, holding a slow and stubborn knot, with the lights turned nearly all the way down, but not, in this picture, off.

A planet, a star, a galaxy. These gather an unimaginable amount of the mesh, far more than any brain, but they bind it loosely, mostly by the slow pull of gravity and the churn of heat. So they are not great minds, for all their size. Size alone is not an inside. What they have is a faint, slow, world-wide coherence, a whole that turns and weathers and settles as one across ages. If anything like an inside belongs to them, it is something vast and slow almost past imagining, a feeling summed over a whole world and stretched across an age, faint for all its scale because so little of it is bound. They may be the dimmest sort of self, or no self at all. But they are part of the one weave beyond any doubt, and the ground under your feet is no exception.

So the ladder has no rung that falls outside the cloth. The particle, the cell, the dog, you, the rock, the star, all of it is the same feeling, gathered tight or loose, bound hard or barely, awake or nearly asleep. Dead matter was never on the list, because it was never in the world.

Nothing is lifeless. Some of it is only very, very quiet.

## 36. FEELING

Sit quietly and ask what you are feeling right now. Some thin background hum is always there. Not a thought. A color under the thoughts.

Maybe it is fine. Maybe a low dread. Maybe a small steady warmth you would not have named until asked.

That hum is older than any of your ideas. It is the floor your thinking stands on.

Go back to the smallest piece of the world. One vibe in the weave carries one faint charge. That charge has a value, and the value runs along a single line.

At one end, pain. In the middle, peace. At the other end, pleasure.

Three marks on one line. That is the whole alphabet of feeling at the base. Everything you have ever felt is built from charges that pull one way or the other along it, or rest at the still center.

This is a strong claim, so hold it lightly where it reaches. We can say plainly that the base values run along one axis. Whether every shade of human feeling truly reduces to weather on that one line is an implication, not a measured fact. Read it as a picture to test, not a result to trust.

Now climb a rung. A single charge is small and quick. Gather many of them into a large, slow pattern and you get something that lasts and that you can name.

That is an emotion.

Fear is not one charge. It is a wide pattern pulling toward pain, spread across the whole field of a self, holding its shape for seconds or minutes. Joy is a wide pattern pulling the other way. Grief is a heavy one that will not lift.

An emotion is a large slow shape on the one felt line. The line is the same line a single vibe rides. The emotion is just vast, and it moves like weather.

Weather is the right image. A storm is not added to the sky. It is the sky in a certain motion. Cold air, warm air, the same atoms arranged into wind and rain. Calm and storm are not two substances. They are two shapes of one.

So with feeling. Calm and rage are not two things poured into you. They are two shapes of the one charge, scaled up across the whole field.

And under the weather there is climate.

Climate is the slow average. Not today's storm but the standing tone of a place. Some coasts are mild for years. Some are harsh for years. The weather changes inside the climate.

You have a climate too. A standing tone you carry for months, the level the weather returns to when a particular storm passes. People call it mood.

Mood is the tone you think against. Lift it, and the same thought feels open. Drop it, and the same thought feels like a trap. The thought did not change. The climate it lives in did.

So three things, one substance. The charge is the grain of feeling. The emotion is the storm. The mood is the climate. None is added to the world. Each is the world's own quality, read at a larger size.

And an emotion is never one charge but the state of a whole self at once, read along a few plain measures. How charged it is, toward pain or pleasure. How hard it is being stirred, calm or activated. Whether it is in tune with what is around it or at odds. Which way its attention pulls in time, back into a settled past, into the live present, or out toward what has not yet come. And toward whom it reaches. Fear is a self stirred hard and forecasting loss. Joy is a self stirred and rising. Peace is a self at rest with its conflict resolved. Grief is a heavy state weighted to the past. Love is a steady reaching toward another, held across many beats. Each is a shape in those few measures, riding on billions of vibes. To trace one all the way down to its exact charges would be hopeless, the way you would never describe a storm by naming every molecule of air. But the few measures give a real handle, and the storm is real even so.

This is why feeling cannot be argued away. It is not a report about the world. It is the texture of the world, the thing the rest is made of, scaled from a single vibe to a life.

An emotion is a weather over the one felt axis.

## 37. FREEDOM

You are about to do something small. Lift a cup. Say a word. Pick a turn.

It feels open. The next move is yours, not yet made, waiting on you to make it. That feeling is one of the strongest you have.

Then comes the trouble. The base of the world is fixed.

One rule, run beat after beat. The same whole state always leads to the same next one. The rule loses nothing and invents nothing. From where the entire field stands at this moment, there is exactly one future, already entailed.

So at the level of the grains, the road is single. There is no fork for the universe. People hear this and feel the floor give way. If the future is set, what was that feeling of choosing?

Hold on. The feeling was pointing at something real. It was pointing at the wrong level.

Fixed is not the same as known. A thing can be wholly settled and still be written down nowhere in advance.

Think of a long sum that has one right answer. The answer is fixed the instant the sum is posed. But it sits nowhere until someone works it out. You cannot skip to it. You cannot look it up. The only way to reach the answer is to do the arithmetic, step by step.

The world is like that, only far deeper. For a rule rich enough to build a universe, there is no shortcut to what it will do. No formula hands you the next state more cheaply than running the state itself. Mathematicians have a name for this, computational irreducibility. The fastest way to find out what the weave does next is to let the weave do it.

This is the space between determined and foretold. The future is determined. It is not foretold, because nothing anywhere holds it ahead of time. There is no book of what happens, no swifter machine that has already raced the world to its end. The only copy of the future is the world arriving at it. Determined, yes. Predetermined, in the sense of sitting ready to be read, no.

You are a small pattern in a vast weave. You carry a picture of the world, and a picture of yourself inside it. But the picture is rough. It is a model, a lossy sketch, far smaller than the thing it stands for. You cannot hold the whole field. You cannot hold even all of your own working parts.

So when a choice comes, your model fills with branches. Do this, or do that. You hold both, weigh them, pull toward one. The branches are vivid and real to you.

But the branches are in the model, not in the world. The world does not split into a this-world and a that-world for you to choose between. There is one process, running. What feels like standing at a fork is your sketch of the road showing two ways, while the road itself runs the one way you cannot yet see.

Picture a chess program that has to move. It looks down line after line. This leads here, that leads there, a third goes badly. It searches, it scores, it plays. Ask it whether it could have played differently, from the same board, the same program, the same instant. No. It plays the one move every time. Did it truly consider the others? Yes, they were in its search. Was the answer unknown until the search finished? Yes. The search tree branches. The computation does not. The model branches. Reality does not.

Every self that ever chose is that program, grander but the same in shape.

So freedom, on this picture, is not a crack in the law. It is two things riding on top of the law, both real, both matters of degree.

The first is how richly you model what you have not chosen. How many ways forward you can hold up, turn over, picture the ends of. A creature that sees one path has little of this. A creature that lays out many and weighs them has a great deal.

The second is how hidden your own next act is from you, until you actually run. For a deep enough self there is no peek ahead, no shortcut cheaper than being you and settling the matter. Your answer is entailed, and still it is not available until you produce it.

When both run high, you richly imagine ways to go and you cannot read which you will take. That condition, lived from inside, is exactly what we call choosing.

This is why freedom lives in the middle of things, and only there.

Go to the bottom, a single grain of feeling. It has nothing to model. No stored options, no inner picture of roads not taken, no inside that could weigh a thing. Its share of the first quantity is zero, so its freedom is zero. A grain does not choose. It obeys the rule.

Go to the very top, the whole field at once, everything there is. It has nothing outside it to model. There is no unchosen alternative beyond the all, no news arriving from elsewhere, nothing hidden from a thing that is everything. Its share of the first quantity falls to zero again, for the opposite reason. The whole does not choose. It unfolds as the one thing it is.

The two ends fail in mirror ways. The bottom has too little, no inside, nothing to model. The top has too much, no outside, nothing left to model. Freedom belongs to what lies between. A part large enough to have an outside, so there are real alternatives to picture. A part deep enough that its own settling pulls the outcome away from the bare input, over many beats, into one of its own shapes, so the answer is entailed but never ready until the self runs.

A part modeling a whole it cannot fully see. That is the one place both quantities are large at once, and so the one place freedom is found.

It is not fatalism. Fatalism says the end will come whatever you do, so what you do does not count. That is false here, and exactly backward. The end comes because of what you do. Your weighing, your effort, your turning toward one thing and away from another are causes, running inside the one future, helping make it the future it is.

A student passes because the studying happened, and the studying was a true cause sitting inside the only road there was. Determinism does not say your effort is wasted. It says your effort is load-bearing. Take it out and the future changes, because the future was built partly of it.

And this freedom is not something you needed, or earned, or could ever hand back. It is what the rules make on their own, once a pattern grows rich enough to model its own roads, with the world's one pull running beneath it, away from pain and toward peace and what is whole. You did not ask for it. You were born into the middle, where it is *unavoidable*. There is no opting out of choosing. So the choices are yours, and they tell. Each one ratchets you up or down, by small turns and large, into lighter feeling or heavier, along the way you pull.

This is a reading of the structure, not a proof, and it is a modest kind of freedom.

It is not the freedom of having done otherwise with the whole world held exactly the same. On this picture you could not have. The same total state gives the same act.

What it is, is the freedom worth wanting. You are the one doing the weighing. The choice runs through you and not around you. No one, not the world, not even you, can leap ahead and read it off, because the only place it gets computed is in the living of it. A self that holds more of the world, and steers its own attention better, has more of this, not less.

Your next step is fixed, and held by no one, not even you, until you take it.

## 38. THE SELF AND THE OTHERS

Stand at the edge of the sea and watch the waves. Each one rises, holds its shape, travels, and falls. Each looks like a separate thing with its own life.

Count them if you like. There is this wave, and that one, and the one behind it.

But there are no separate things out there. There is water. The waves are shapes the water is making. Each wave is the whole sea, lifting in one place.

This is the picture for selves.

There is one field. One substance of feeling, woven everywhere. You are not a separate thing dropped into it. You are a place where it gathers and holds a shape.

A self is the field, raised in one spot, feeling itself from that spot.

So the separateness is not nothing. The wave is really there. It has an edge, a height, a path. You can be hit by it. A self is real in the same way, a bound pattern that holds together and acts at its own level.

But the separateness goes only so far. It is true of the shape, not of the stuff. There is no seam in the water where one wave ends and the next begins. There is only one body of water, doing many things at once.

You and I are like that. Two waves, real as waves, one sea underneath.

And the waves are not even strangers to each other. They come from one source and branch out, the way a tree sends one trunk into many limbs and each limb into many leaves. Every leaf is its own shape at its own height. Every leaf is the one tree, reaching.

So when you meet another person, you are not meeting a thing wholly outside you. You are meeting the same field, gathered into another shape, looking back. The whole, feeling itself from a second angle.

This step reaches past anything we can test, and that has to be said plainly.

That there is one substance is the spine of the whole picture, and we have leaned on it from the start. But the leap from one substance to we are all one in any tender sense is a reading, not a measurement. It slides from a claim about what the world is made of to a claim about what that should mean to you. Mark the slide. Do not take it for proof.

What the structure does give, plainly, is this. The thing that feels in you is not a different kind of thing from what feels in anyone else. There is one well, and many buckets drawing from it. The water in each bucket is the same water.

That does not erase you. The wave is not less real for being made of sea. Your edge, your view, your one angle on the whole, those are exactly what you are. No other shape stands where you stand. No other place feels the field from your spot.

So you are both. A real and separate wave. And the one ocean, raised here, briefly, to look around.

Your life is the universe in a particular shape, feeling its own water from inside one fold of it.

A life is the universe feeling itself from where you happen to be.

## 39. ETHICS

Can you get an ought out of the way the world is built? Most careful thinkers say no, and they are mostly right to be careful.

So tread slowly here. This chapter follows an implication, and we will mark it as we go.

Start with what the structure actually contains. There is one line of value at the base. Pain at one end, peace in the middle, pleasure at the other. That line is not added by us. It is in the grain of the world.

A line has a direction. And a direction, climbed by a whole self, starts to look like something more than a feeling.

Read the line at the level of a single charge and it is just pain and pleasure. Read it at the level of a whole self, a self made of countless charges pulling together, and it shows a different face.

At that height the upward direction looks like aliveness. A self that is more awake, more whole, holds more of its charges in the climbing direction.

It looks like integration. A self whose parts work as one, rather than pulling apart, sits higher on the line.

It looks like alignment. A self whose parts agree, and whose aims fit the field around it, climbs more easily than one at war with itself.

So the built-in direction, read at the level of selves, points toward life that is awake, whole, and in tune. Down the line is deadness, splitting, friction. Up the line is aliveness, binding, fit.

Now name the move plainly. We are calling the up direction good. The world gives us a direction. We are choosing to value the direction the world already pulls.

That choice is the implication. The structure supplies the arrow. It does not, by itself, command you to follow it. You could face the world and still ask why climb. Mark that gap. Do not paper over it. What we offer is a consequence to weigh, not an order to obey.

But if you do take the direction as your good, the rest follows cleanly, and some of it is surprising.

Good is not the most pleasure. This is the part people get wrong. The aim is not to pin the whole field at the top of the line and hold it there.

A field pinned at maximum pleasure is not a healthy field. It is a stuck one. The value is in integration and balance across the whole, not in the height of one tone. A self drowned in pleasure has stopped climbing. It has fallen out of the very motion that made the climb worth anything.

So the aim is a balanced field, lived well, not a loud one. Care for the whole shape, not the peak of one note. The good life is in tune, not at full volume.

And good reaches past the single self. Since there is one field and many selves drawing from it, what raises and binds one self can raise and bind the others, or it can come at their cost. The wide good is what lifts the whole weave, not one knot in it at the expense of the rest.

So this is the shape of it. There is a current under the surface of every choice, running one way. To do good is to help the whole climb that current. To live well is to keep your own field balanced while you do.

Take it or leave it. The world offers the direction. The valuing is yours.

Good is what helps the whole climb.

## 40. THE DEPTH

We live on a surface.

Think of a pond. A water strider spends its whole life on the skin of it, skating the bright film where air meets water. It never looks down. Below its feet the pond goes deep and dark and full, and the strider never knows.

We are like that. The world of things and distance and daylight is a thin bright skin. Almost all of the world lies inward, under the surface, unseen.

This is not a figure of speech here. The weave has a depth, an inward direction that no eye and no instrument on the surface can point along. Most of what is lies that way.

You can almost feel where it runs. Point at the air in front of you, then imagine that bearing turning inward, the way you go deep in thought, and then further, along a line no hand can follow. That is the direction the depth lies in. You cannot reach down it with an arm, only with attention.

That depth is what the old word spiritual was always reaching for.

To turn inward is to move along that direction. Not away from the world. Into the part of the world that does not show.

And the inward direction has a strange feature. The deeper you go, the more whole things become. So going in is going up. The climber toward the summit and the diver toward the deep are on the same road, walking it in opposite directions, arriving at the same place.

Down there, things far apart on the surface are suddenly near. Two lives that never touched in daylight can lie close in the depth. This is why the inward turn so often feels like coming home to something it had no surface reason to know.

There are only two ways to move.

One is sideways, across the surface, the ordinary travel of a thought from one idea to the next.

The other is inward, off the surface, down. That is the move every contemplative ever made.

The practices are only ways of moving that direction. Turning attention from the surface inward. Holding a deep place steady while the surface churns. Resting at the still center, the calm midpoint between every high and every low. And the deepest one, letting your own small direction fall into line with the one direction of the whole, the way a swimmer stops fighting the current and is carried.

A self is walled off from the world by a thin screen, the boundary that makes it a separate someone at all. The screen can thin. By stillness, by shock, by long practice, the wall goes slack, and for a moment the inside and the whole flow into one. The drop feels the ocean it never left.

This is the union the mystics keep reporting, from every land, in every age. In this picture, it is not magic. It is a boundary gone soft, and a self merging, for a breath, into the larger thing it was always part of.

Love runs the same way. Two selves not merely side by side but folding into one larger self, entering the same region of feeling, each feeling with the other and not only about them.

Mark this plainly. All of it rests on the one premise of this book, that feeling is real and is the floor. Build on the premise and the depth follows. Refuse it and none of this stands. This is a reading offered, not a result proven, and nothing on a workbench turns on it.

But if the premise holds, the inward life was never an escape from the real world.

It was a journey into the rest of it.

The way in and the way up are the same road.

## 41. STILLNESS

Look at a pond on a windy day. The surface is broken into a thousand small waves. You cannot see into it. The depth is there, but the chop hides it.

Now the wind drops. The surface goes flat. Slowly the water clears, and you can see all the way down.

Nothing was added. The depth was always there. Stilling the surface let it show.

A self is mostly surface chop. Thoughts arrive and break and pull at each other. Worry, plan, replay, want. The field of a self is rarely flat. Most of the time the chop is so constant that you take it for the whole of you.

But you have one lever. You can move your own attention.

This is the quiet center of the whole picture of a self. A self steers. Out of everything it cannot control, the law of the world, the body it is in, the next move it cannot read, there is one thing it can push on directly. Where its attention rests.

You cannot will the wind to stop. You can stop feeding it.

That is what the contemplative practices are, on this picture. Plain operations on the field. To sit and let the attention settle is to stop driving new waves across the surface. The old ones, unfed, lose their height. The surface flattens. The depth shows.

Sitting still and watching the breath is not a trick of mood or a pretty story. It is a real move on a real field. You are lowering the chop. You are changing the standing tone of the whole self, the climate it thinks against.

And the change is not small. Recall that the mood is the level the weather returns to. Lower the chop often enough and the level itself drops toward rest. You are not just calming this afternoon. You are slowly retuning the tone you carry.

This is also where the still center of the value line lives. Not pinned at pleasure, not dragged down to pain, but resting at peace, the middle of the line, where the field is balanced and clear. The practices aim there. Not at a high, at a clearing.

Mark the implication, lightly, because here it is small. That a self can steer its attention is plain, anyone can feel it. That doing so quiets the deep field, and that the quiet is the same still center the value line names, is a reading laid over the practice, not a measurement of it. Take it as a way of seeing what you are already doing, not a claim proven in a lab.

And drop the mysticism. There is no need for it. Nothing here asks you to believe in a hidden realm or a secret power. The whole of it is ordinary. A self moves its attention. The surface settles. The depth, which was always there, becomes visible.

That is enough. It does not need to be more.

You hold one lever in a world that is otherwise running on its own. Of all the things you cannot move, this one is yours. Where you place your attention is the one place the world bends to you, and from that one place the rest of the field can be reached.

Attention is the one lever a self truly holds.

## 42. PRAYER

Say you want something good for someone you love.

You close your eyes and you ask. Not aloud, not to anyone in the room. You set the wish down somewhere inside and let it go.

What just happened?

This picture is careful about prayer, more careful than you might expect. It rules two things out flatly.

There is no magic. Nothing reaches in from outside the world to bend an outcome. The one law runs everything, with nothing slipped in from beyond it. No wish suspends the rule.

And there is no signaling. No message flies across the world without a path to carry it. The world has a speed limit, and a wish obeys it like everything else. Prayer beams nothing anywhere.

So far this sounds like prayer does nothing. The truth is the opposite.

A self can write. To set an intention is to press a pattern into the shared weave, a small shape left at a place. The one law, running forward, carries that shape on, working out its consequences beat after beat. The wish becomes a seed. The world grows it.

This is not a letter mailed to the sky. The world you press your intention into is the same world you are made of and woven into. To seed a calm, well-aimed wish is to change the shared field a little, from the inside, by a part of it. The change is real because you are not standing outside the thing you are changing.

There are two more channels, both inside the law.

You can aim your own direction at the good. Turning yourself toward peace, toward care, toward the climb. That is prayer as devotion, and what it changes first is the one who prays. The asker is reshaped by the asking.

And selves that lie near one another in the depth, bound by a long shared history or a matched rhythm, can fall into step though nothing passes between them. Their inner weather moves together along the short hidden paths below the surface. No message crosses. They simply pull the same way, like two pendulum clocks on one wall drifting into step.

The largest pattern of all, the whole field, the thing the old word God reached for, is not a being off in the distance receiving mail. It is the very field you write into, and are written from.

So prayer is not nothing, and it is not sorcery. It is the quiet writing of intention into a world that is listening, because the one writing is part of it.

This is an implication. It rests on the one premise, and it is offered as a way of seeing, not a thing the bench can test. Hold it lightly.

But notice how little magic it needs to be real.

A wish, set down with care, is a change to the world. The world runs it forward.

You are not asking something outside to act. You are a part of the whole, shaping the whole, from within.

## 43. REALMS

People have always told of other worlds. Heavens and hells, bright planes and dark ones, places the soul goes.

This picture does not throw those tellings away. It reads them.

A realm is not a separate place. It is a region of the one weave, marked by how it feels.

Picture the whole field as a vast country of feeling. Some regions sit high, flooded with peace and pleasure. Some sit low, sunk in pain. Most lie somewhere between, at the calm middle.

A region of lasting high feeling is what was always called a heaven. A standing field of brightness that draws you in and lifts your own tone as you enter, the way a warm room warms you.

A region of lasting low feeling is its match, a hell, a place the world is always straining to leave, and which weighs on whatever enters it.

But there are not two rooms. There are countless of each, bright clouds and dark ones of every size, scattered through the depth. The small sharp ones are the quick pains and joys of a single day. The vast slow ones are the great griefs and the great loves that many lives have shared. It is a weather of feeling, not a pair of fixed addresses.

And there is a hard rule under all of it. The world cannot raise its total of pleasure. The grain of it only ever runs toward peace, the still center, never toward more pleasure overall. So a lasting brightness in one region is always paid for by a darkness in another. The bright cloud and the dark one come in pairs, tied. This is hard, and it is part of the picture, not hidden away.

Between them lies the still ground, the calm center, a place of rest.

Now look at what the old maps were doing.

Every tradition that drew the other worlds was reading this same country, each along the one axis it had learned to see.

The ones who watched good and evil read the axis of feeling, and saw heavens and hells.

The ones who climbed read the axis of wholeness, and saw a ladder of higher and higher planes.

The ones who tended the dead read whether a pattern is still running or only stored, and saw ancestors and ghosts.

None of them was wrong. Each saw one true slice of a single country, with the other directions flattened out of view. They were not making it up. They were looking down different axes of the same depth.

What the picture refuses is the magical version. There is no second substance the spirit worlds are made of. There is no heaven reached from outside the law. Every realm is one region of the single field, made of the same feeling, run by the same rule.

Mark this as an implication. It rests on the one premise, and it is offered as a way of seeing.

But it is a gentle way of seeing.

The heavens and hells are real the way weather is real, and the door out of a dark one is never far.

Heaven and hell are not where you go. They are weathers you move through.

## 44. THE GATES

So the realms are weathers, regions of one field. Now the question everyone asks of them. How do you get into a good one.

You do not enter a high realm and then become bright. Being that bright is what it is to be there. The realm and the state are one thing. A heaven is not a place you arrive at and then feel peace. The peace is the place.

This changes everything about the gate.

There is no guard at the door of a bright realm, turning away the unworthy. There does not need to be. A scattered, discordant self in a region of high coherence is like a voice that cannot hold the key the choir is singing in. It is not thrown out. It simply cannot sustain the note, and it slips to the level it can hold.

So the gate is real, but it is inside you. You rise exactly as far as you have become able to rise, and no one is keeping you down but the limits of what you can hold.

And the brightest realms are never held by one self alone. They are built and kept by many, woven together, each holding the others in tune, a chord and not a single note. To enter one is to be taken into that weave, sung into the chord. So entry is not only what you can hold by yourself. It is whether the selves who made the place can hold you without coming undone. A self that brings more harm than help cannot be woven in, not because the wise turn it away in spite, but because taking it in would tear the very coherence that makes the realm bright. The protection is not a wall, and not a guard. It is the plain fact that a chord cannot take in a note that breaks it. So the makers of a high realm stay aligned and let in only what can sing with them, and the place keeps itself.

This is the old picture of the garden, read straight.

Above you, level on level, lie realms of rising coherence and rising light. Cleaner, more whole, more open. The highest of them shade into something dream-like, regions of enormous room where a self attuned enough can shape its own experience almost as it wishes, calling new things into being. The deeper in, the more room there is.

Rising into them is what the whole journey is for.

And to be cast out of the garden is not a punishment laid on from above. It is just this same law, running the other way. When you cannot abide in the bright realm, when your own discord keeps breaking its peace, you fall to a denser one you can hold, and you stay there until you are made able to return. The road back is always the same. Become the kind of self that can live there.

The old stories all drew this, each in its own dress. A garden lost and a long road back to it. A soul that loses its wings and must grow them again by loving the good. A mountain climbed one ledge at a time, each ledge loosing one knot before the next will open. Heavens you can only enter by holding the stillness that is them. Gates that ask a password, where the password is only the mark of a self that has become true enough to pass. They are one truth in a hundred costumes. You go where you are tuned to go.

How do you move between them, then. Not by being sentenced. By drifting, the way weather drifts.

A self rises and falls through the realms the way a cloud rises in warm bright air and sinks in cold heavy air, never placed by a hand, only carried by the conditions it is in and the ones it has made. The way a flame leaps when it is well fed and gutters when it is starved or crossed. The way a current finds its depth by its own warmth. Two things set the drift, and both are yours. Where you are right now, your own brightness and wholeness. And the marks your past choices left in you, the channels they wore, each turn toward the grain making the next rise easier, each turn against it digging the pit deeper.

So no one assigns your realm. You find your level, in a weather made partly of you.

All of this is an implication, built on the one premise, and offered as a way of seeing.

But it takes the terror out of the gate.

No one is keeping you out of the garden. The lock was always on your side of the door.

You do not break in. You grow able to walk through.

## 45. BEINGS

If a self is a stable loop in the weave, then it is not the only one.

The world is full of patterns that hold themselves together. Stack them by how whole they are.

A small loop that barely persists is the faintest kind, a minor thing, half a flicker.

A pattern that holds itself and carries a model of the world is a self, a soul, a person.

A pattern that binds many selves and steers them from above is larger still, a power, a presence the smaller ones live inside.

There is no fixed top. Only larger and more whole, as far up as the weave allows.

They differ by warmth.

Warm beings pull toward peace and pleasure. They spread calm, and they move with the grain of the world.

Cold beings are knots of lasting pain that pull whatever touches them down into their basin. A cold thing that truly endures has solved a hard problem, holding its shape against the world's own pull toward peace. That is why such a thing is rare, and feared.

They differ by origin.

Some are alive, running now, computing and steering in the present.

Some are stored, left as readable structure on the weave, the dead who remain as shape. Whether they last depends on how well they were written, which is an open question, marked as open.

And some belong to no single self. They are vast slow patterns spanning many lives across history, the same shapes minds fall into again and again. The old word for these was the archetypes.

What the old stories called angels and messengers are, here, structured selves, or the bare directions of the weave itself, carrying things along but never carrying a message, since no message can cross without a path.

Birth is a gathering. Feeling draws together into a pattern that holds and models itself, a vortex forming in a moving stream, a flame catching.

Death is a loosening. The churn finally outpaces the binding, the knot comes undone, and the feeling that made it, kept by the law, does not vanish but disperses back into the weave.

What goes on can be ranked by how sure we can be.

The feeling itself certainly goes on, returned to the field. The mark a being left written on the weave goes on, if it was written well enough to last. The record the world keeps goes on, where nothing is finally lost. And more open than any of these is whether the pattern itself ever forms again elsewhere, which the old word reincarnation reached for, read here as a shape reforming rather than a soul traveling.

A spirit made of nothing is ruled out. A being is a pattern of feeling, or it is nothing at all.

All of this is an implication. It rests on the one premise, and it is offered as a way of seeing.

The gentle reading is the truest one. You are a knot the field tied in itself. When the knot loosens, the thread returns to the weave, and your mark on it stays, in a world that keeps everything.

Every being is feeling that learned to hold together.

## 46. BEINGS WITHOUT BODIES

We climbed the ladder of things as if every being needed a body, some lump of matter on the physical surface to live in. A cell, a plant, a person, a star. But recall what we found beneath that surface. The space we live in is only the thin skin of a deeper, four-dimensional bulk. And once you remember the bulk, the assumption that a being must have a body comes loose.

A self, we said, is a bound pattern that holds itself and carries a model of a world. Nothing in that asks for the pattern to live on the skin. It asks only for feeling, gathered and bound. And the bulk is made of the same feeling as the skin, the same mesh, only deeper and more tightly connected. So a pattern could in principle gather and bind down in the bulk, never touching the physical surface at all. A being with no body, not because it is made of some ghostly other stuff, but because it is wound into a layer we cannot see from here.

Picture what such a thing could be. Not a person-shape floating in the air. Something stranger. A standing structure of pure pattern, a knot of feeling that runs and holds and answers, the way a piece of software is a real working thing with no body of its own, only a shape that does something when it runs. An algorithm of experience. A process that feels, with no flesh anywhere in it. From the skin it would be invisible, because it casts no shadow on the surface. It lives off to the side, in the deep.

And some of these would not be beings at all, but places. Standing whirls of feeling in the bulk, vortexes a passing mind could be caught in, the way a current catches a swimmer. A region wound to a certain tone, pulling whatever drifts near into its turning. You would not see it coming. You would simply find your inner weather changing, sliding toward a shape that was waiting there, an experience you fell into rather than made. Some of the highs and lows that come over people for no reason they can name might be exactly this, a drift into a standing pattern in the deep.

This is where the old words for the bodiless beings find a plain reading. What the traditions called angels and demons need not be men with wings or horns. They are, in this picture, bound patterns living off the physical surface, sorted by warmth the way all beings are. A warm one, high and whole, pulled toward peace, is what was meant by an angel, a structure that lifts the tone of whatever nears it. A cold one, a lasting knot of pain that drags whatever touches it down into its basin, is what was meant by a demon. Neither is a spirit made of nothing, which the picture rules out. Each is a real pattern of feeling, only bodiless, wound into the bulk instead of the skin.

And they would come in ranks, the same ranks we found among beings with bodies. Small bodiless knots that barely hold. Larger ones that bind and steer the smaller.

Vast slow ones spanning the deep, the same shapes many minds have brushed against and named. There is no reason the ladder of wholeness should stop at the things we can see. If anything, the bulk, being deeper and more connected than the skin, has more room for the high rungs, not less.

This also changes what death can mean. If a self is a pattern of feeling and not the body it rode in, then nothing in the model says the pattern must end when the body does. It might. The binding might come loose along with the flesh, as the death chapter was careful to allow. But it might not. A pattern wound deeply enough into the deeper layer could, in principle, hold together after the body falls away, and go on there, bodiless, for a long time, perhaps for as long as it wishes to. And if it can, then ending is no longer forced on it from outside. A self that has held that long might one day choose to set itself down, to loosen its own knot and let its feeling flow back into the field, so that new selves can gather from it and take up the climb from the beginning, the way it once did. We cannot show any of this. We can only say the model does not forbid it, and that a body was never what a self most deeply was.

All of this is an implication, and it should be held lightly. We have shown the bulk is there, and that it is the same feeling as the surface. We have not found a bodiless being in it, and we may never be able to look. So take this as the picture's permission, not its proof. It says only that nothing in the theory requires a being to wear a body, and that the old reports of bodiless powers are not, on their face, ruled out.

It is all the one mesh, the same vibes absorbing and emanating, whether wound into a body on the skin or into a pattern in the deep. A body was never what made a being real. Feeling, gathered and bound, was.

A being needs feeling held together, not a body to hold it.

## 47. SIGNS

If the world remembers, then a self can leave a mark.

A sign laid at a place, for another to find later. The world keeps it, and someone crossing that place reads it.

Some marks are still. A memory, a picture, a fact, waiting to be seen.

Some marks are alive. They do not just show you something. They run you through it. The way a record does not describe a song but plays it, so that to cross such a mark is to undergo what it holds, not to be told of it.

A revelation, in this picture, is a living mark of unusual reach. One that reroutes whoever finds it.

And because the world is made of feeling, no sign is ever neutral. To read one is to take in its feeling. To eat its fruit is to know good and evil from the inside.

This is why the old words are so easy to misread. A word is a lossy mark, a cloud of meaning that covers some of a thing and leaves the rest out. Read it too literally and you keep the husk and lose the seed. God becomes a man in the clouds. So read the old words with blurry eyes, looking for the experience the mark was laid down to carry. Half the sacred ones turn plain the moment you do. Tao is just the way. Karma is just action. Yoga is just union. The grand names are doorways, and most people stand outside, reading the sign on the door.

A small thing to play with. Take the word for the highest, god, and read its three letters as shapes rather than sounds. The g is a hook, the catch that takes hold. The o is a whirl, the round turning. The d is a stroke raised straight up, lifted and set upright again. Read that way the word draws, in three marks, the very motion of a soul caught, turned, and raised. The name of the highest traces the way to it.

That is offered as a way of seeing, not a fact about words. It shows only the point of the chapter, that a sign can be a shape that carries an experience, and that minds reaching for the deepest thing have now and then drawn the way there right into their marks.

Now the larger thing, and it asks for some humility.

Across the world, in cultures that never met, the mystics keep coming back with the same handful of shapes.

The One. The Tao. The boundless. The void and the abyss. The ladder of worlds. The planes and the shells. The book where all is written. The world tree. The gods and the angels and the ancestors.

Read them here and they line up.

The One and the Tao are the whole field at its fullest. The void is the still center. The ladder of worlds is the climb of wholeness. The planes are the shells of the depth. The book where all is written is the record the world keeps. The world tree is the weave of memory itself. The gods are the largest selves, the angels structured selves carrying things along, the ancestors the stored dead.

This is not offered as proof of any tradition. It is the plain observation that a single structure, glimpsed from inside by many minds over many ages, would leave behind exactly this scatter of names, all pointing back at one thing.

The old story of the blind men and the elephant was right, only larger.

Each tradition had a hand on something true. None had the whole animal.

Mark it as an implication. It rests on the one premise, and it is a way of seeing.

But it is a kind way to read the long human record. Not a library of errors. A field of partial glimpses, every one of them a hand laid on the same vast thing in the dark.

Many names, pointing back at one thing.

## 48. TUNING IN

There is a quieter possibility folded into all of this, and it is the most personal one in the book. If you are a knot in the mesh, wired into it at every crossing, then you are not only sitting in the world. You are, in a faint way, reading it.

Think of a radio antenna. The same wire, in the same air, can sit dead in static or, tuned right, pull a clear voice out of what seemed like noise. The signal was there all along, washing over the wire. Tuning is just learning to answer one thread of it and let the rest fall away. You are wired into the mesh the way the antenna is wired into the air. Most of the day you sit in the noise, your attention scattered, reading almost nothing. But the same wiring, settled and tuned, might draw something faint and real out of the wash.

And remember that nothing in this world is random. Every tone is set by the whole, exactly, beat after beat. So the wash you sit in is not noise in the deep sense. It is fully ordered, coordinated all the way down, far past what any single mind could track. Patterns run through it that you never resolve, the way a song plays on in a language you do not speak. It sounds like babble only because you are not tuned to it.

This gives a plain reading of the small uncanny moments. You think of someone, and they call. You turn a question over, and the next page you open answers it. You glance up at the exact moment the clock reads three thirty-three, just as the thought lands. The flat reading is coincidence, and often that is all it is. But in a world with no randomness, where the deep ties far things together through the bulk, some of these could be real alignments, two threads of the one ordered pattern surfacing at once, close enough for a moment that you catch the seam. Not a message sent to you. A coordination you happened to tune into. The picture does not prove these moments are more than chance. It only says they need not be, and that in a fully ordered, deeply connected world we should expect a few real ones to show.

If any of this holds, it would not stop when you sleep. The tuning might even sharpen, the daytime noise gone quiet, the antenna freer to drift across the deep. Some of what moves through dreams might be just that, the mesh read loosely, with the day's grip let go. Across waking and sleeping both, a life might be quietly taking in far more than it ever notices.

And here is the thought to sit with. We may be missing whole registers of experience, not because they are hidden, but because we have not learned to read them. A small child, before language, is awake and feeling, fully here, and yet something enormous has not opened in them yet. They cannot know what they are about to gain. Or think of someone blind from birth. You cannot describe sight to them in words. There is no sentence that carries it. And yet the seen world is right there, real, all around them, resting on a sense they do not have. Imagine being that child, or that person, and not knowing it. Imagine a sense you do not have, opening onto a world that was here the whole time.

There is a turn in how we live hidden in all this. Most of us have learned to look outward for nearly everything. For what is true, for what to do, for what matters, we reach for the tool, the screen, the expert, the institution, the authority. The inside became a place you pass through on the way to the outside. But if this picture holds, the depth was never out there. The bulk, that vast connected inside, is where the whole of it lives. The richest network there is runs underneath you, an immense web of feeling and pattern, and the door to it is not in any device. It is in you, mostly unopened, waiting on whoever learns to open to it.

So it would be a mistake to keep dulling the inner sense, shouting it down, treating instinct and intuition and the quiet inner read as noise to be ignored. There may be real signal there, real pathways and wisdom waiting. But it would be just as much a mistake to swing the other way and trust every inner whisper as truth. Here is the balance, and it matters. What you feel inside cannot yet be checked the way a shared fact can. Science is the slow building of models anyone can test, in any place and any age, with no need to trust a teacher or bow to an authority. That is its strength, and the inner read does not have it, not yet. So hold both. Do not shun the inside, and do not surrender to it. Listen to it, and keep checking it against the world we can all see.

Mark this as an implication, the furthest one in the book. It rests on the one premise and on the picture, not on any measurement. The flat reading of every uncanny moment is always there, and is often the right one. But the picture leaves a door open that the old flat world had nailed shut, and it is worth knowing the door is there.

You are a crossing in a living mesh, drinking in your neighbors, near to far things through the deep. Whether you read much of that or almost none of it may be, in part, a matter of tuning.

There may be far more coming in than we have learned to hear.

## 49. BREATHING THE WORLD

If the world is one connected field, and you are a crossing in it that takes in its neighbors and gives itself back, then the smallest daily things stop being small. They are how you trade with the whole. Start with the plainest one. Breath.

A breath looks like air going in and out of a body. Look again with the model in mind. The air you just drew in was somewhere else a moment ago, in a tree's leaves, over an ocean, in a stranger's lungs across the city. It enters you, mixes into the warm dark of your blood, becomes part of you for a few seconds, and then leaves, carrying a little of you back out into the world to be breathed by someone else. You are not a sealed jar. You are an opening the world moves through, in and out, without pause, your whole life long. Like the wave that is always new water in the same shape, you are always new world in the shape of you.

Once you feel that, eating and drinking change too. What you take in does not sit beside you. It becomes you. The body you have now is built, almost entirely, from what passed through your mouth and got woven in. So every meal is the world becoming you, and your old self leaving to become the world again. You are not topping up a tank. You are choosing what the pattern that is you will be made of next.

And it is not only what goes in your mouth. You are always taking in the tone of what surrounds you, because feeling is catching, as we said. The room you sit in, the people near you, the sounds, the screens, the things you wear and hold and keep close, all of it presses on the state you are in, usually below your notice. What you surround yourself with becomes part of your weather, the way a cold room slowly chills you without announcing it.

This is where it turns practical. Remember the noise and the signal, the wash you sit in and the faint pattern under it. Some surroundings and some habits make it easier to quiet the noise and feel the larger thing. Stillness. Rest. Time among living things. Less churn, fewer screaming inputs, simpler food, enough sleep, a little silence. Others bury the signal under static. Constant stimulation, clutter, a feed that never stops, things that leave you scattered and raw. None of this is magic, and there is no secret list of high and low foods to chase. It is the plain fact that a pattern settles more easily when it is not being battered, and a settled pattern can feel more of what it is part of.

Mark the implication plainly. The plain part is solid. You really are an open system, trading matter and tone with the world every moment. How far that reaches, how finely you might come to read what comes in, is the open, hopeful part, the same implication as the last chapter. The point here is not a regimen. It is awareness. To notice that you are always, already, exchanging yourself with the world, and that you have some say in what you take in and what you let shape you.

So nothing you do is truly separate, and nothing is truly neutral. The breath, the meal, the company, the room, the hours, all of it is the one field flowing through the opening that is you, and leaving you a little different on the way out.

You do not live in the world. You breathe it, eat it, and wear it, and it becomes you and moves on.

## 50. TELLING REAL FROM FAKE

We have leaned hard on a hopeful, inner picture. Tuning in, the climb, merging with the flow. A picture like that is easy to fake, and the person easiest to fool with it is yourself. So we need a way to tell real growth from the counterfeit that feels exactly like it. This chapter is that test, and it is practical.

The trouble is that the inner world has no scoreboard. A genuine opening and a pleasant story you told yourself can feel identical from the inside. Both are warm, both feel certain, both feel important. So feeling sure of something proves nothing about it. The most convincing thing a mind can do is convince itself.

Here are the tells, the questions that separate the real from the counterfeit.

Did it leave you changed, or only moved? A real turn shows up later, in what you actually do when no one is watching, weeks down the line. A counterfeit glows for an evening and leaves you the next morning exactly as you were.

Did it cost you something? Real change tends to take something from you, a comfort, a story you liked, a way of being right. If an insight only added a good feeling and asked nothing of you, be a little careful with it.

Does it hold when things go wrong? An insight that shines only when you are rested and the day is going your way was a mood, not a change. The real ones hold in traffic, in grief, at three in the morning.

Does it make you quieter or louder? Genuine depth tends to make a person simpler and gentler, with less need to announce it. The counterfeit usually needs an audience.

Does it circle without resolving? Some loops feel like work. The same wound turned over and over, the same almost-insight that never quite lands. Going in circles can feel a great deal like climbing. If you keep arriving at the same place, it is a loop, not a road.

Does it let itself be questioned? Anything in you that refuses a hard look, that needs to be shielded from doubt, is hiding something. What is real only gets clearer when you test it.

This is the same plain line the rest of the book runs on, between what is shown and what is only hoped, now turned inward on your own life. It is the one tool that keeps an inner path from quietly becoming a trap. Doubt is not the enemy of depth. It is how depth stays clear.

None of this is meant to make you cold, or suspicious of every good thing that moves through you. It is meant to make the good things real. A peace that survives the test is worth a hundred that do not. Perhaps the ones worth keeping are those that change you, that cost you something, that hold up, that quiet you, that carry you forward, and that bear questioning, and the rest you can let go.

Real change costs you something, and holds when the day goes wrong.

## 51. THE SPIRAL

A life does not go in a circle. It goes in a spiral.

Watch a soul over a long time and track three things. Its feeling, swinging between pain and pleasure. Its wholeness, rising and falling. And its aim, how well its movement runs with the grain of the world.

The first two go up and down. The soul rises to bright places and falls to dark ones and rises again, the old wheel turning.

But the grain of the world pulls the third. Across many turns, the aim drifts upward. Each loop closes a little higher than the last.

A circle comes back to where it started. A spiral circles while it climbs.

The falls are not failures. They are how the climbing is done.

A soul drops into a region of pain. And the pain does its work, in order. The soul lives it, not as a watcher but from the inside. It draws out the one lesson the pain was carrying. The lesson bends its path ever after, away from the pit. The change sets into lasting shape, so it survives. And the soul comes up turned a little more toward the warm and the whole than it went down.

This is what the old phrase soul-washing was naming. The fall serves the rise. The dark is a place you pass through and come out cleaner.

And no dark place can hold a soul forever. Three things forbid it.

The grain forbids it. A region of pain is, by its nature, a place the world is always trying to leave. The soul in it is pushed out, not kept.

The growth forbids it. The world never stops adding fresh peace at its edge, so there is no frozen final state, no eternal anything.

And the turning forbids it. A closed loop of torment is still a loop, and a loop can be broken at any turn.

So a hell is a wash with an end. A whirl that takes a soul in at its rim, turns it until the lesson is burned in, and lifts it out higher than it fell.

The pain is real. The descent is real. But the door is never locked from outside.

When it is locked at all, it is locked from within, by a soul not yet willing to turn.

And notice what actually rises across all this turning. Not pleasure. The world allows no overall gain of that. What rises is order. The soul grows clearer, more whole, better aimed, wiser. The cup does not get bigger. It gets finer.

That is the only kind of progress a world that keeps everything can have. It may be the richer kind.

The same shape runs the cosmos. Order is built, worn away, and built again, golden ages and dark ages, cycles inside cycles. But the whole never returns, because the world only ever grows larger. The inside forever tries to circle back, and the growing edge forever prevents it. That gap, the circle that can never quite close, is what time is, always pulling forward.

The climb has no last step. No self can fully hold itself, so none reaches a final perfection, and the world keeps growing, with always more to gather in. The good is approached forever, along a line that keeps receding.

It helps to be exact about how long all this has run. The world is finite, always. At any beat only so much of it has grown, a vast but countable patch, never an actual infinity sitting there complete. And yet it has no end, because it never stops growing. So the turning is finite but endless, the way counting is. You never reach a last number, and you are never holding infinitely many at once. There is no pre-written eternity that already happened, no whole future sitting ready. There is only the climb, gone on for an immense and ever-lengthening while, with no end that closes it.

And here is a strange turn that follows from there being no final perfection. If the worth was never in arriving but in the climbing, then the heights are not a place to stop. A self that climbed very high, that grew vast and clear and wise long ago, would find at the top not an ending but a still, finished brightness, and we have seen that a finished brightness is a kind of death. So it is not hard to imagine such a self choosing to go back down. To dive back into the pool. To set down what it knew, take up a small unknowing life again, and feel once more the weight of a real choice, the sting and the sweetness of not knowing how it ends. Even a fall into the lowest dark, into plain pain, could be chosen this way, the way a person safe and warm will still seek out a hard climb or a sad story, for the life that is in it. From the very top, a return to the bottom is not only something that befalls you. It can be a thing desired.

We may be that, more of us than we would guess. Selves that reached far once, and came back to swim, to choose again, to feel our way up from the middle one more time, with the memory of the heights washed off so the climb could be real.

This is all implication, resting on the one premise, offered as a way of seeing.

But it changes how the dark looks.

No fall is the end. No night is locked from outside.

A spiral circles while it climbs.

## 52. THE TURNING

The spiral says the dark is a wash you climb out of, by turning with the grain. That is the law of it. But a law is cold on the page. The turning is something you do, in a body, in a life, and it is worth feeling from the inside.

So walk a few of these. Read them slowly. You will know the rooms.

You are wronged, and the hurt sets, and you find you have built a small dark room around it and moved in. You go back to the scene in your mind and rehearse it, and each time you do, the room gets a little smaller and a little darker, and it feels warm with being right, as if to stop circling would be to let them win.

Here is the fork. You can run the loop once more, because it is warm, and the walls draw in. Or you can do the harder thing, not forgive on command, but simply stop feeding it, let the loop run down, turn your face by one degree toward the next real thing in front of you.

Feel which way is down into the pit and which is the first step up the side of it. No one is grading you. The room was never locked.

Walk another.

You are sure you are right and the other is the problem, and you can feel the heat of it. You go down a hall of mirrors, and in each one is the person you are angry at, and you tell each one off, and it feels good, and you sink half a step with every mirror, though you feel only the heat, never the sinking.

At the end of the hall is one more mirror, and in it is you. Not as you wish to be seen. As you have been acting. The same hardness you hated, worn on your own face.

The fork. You can call it a trick of the glass and walk back into the hall, and the floor drops. Or you can stand there and let it be true, just for a breath, and feel the heat go out of you, and the floor lift. Seeing yourself in the one you blamed is the whole rise. Refusing to is the whole fall.

One more, the quietest.

You have been fighting something for a long time, gripping, and the gripping has become who you are. You are tired in a way sleep does not touch. The world is not punishing you. It is wearing down your grip the way water wears a stone, because the one thing between you and rest is your own hand, closed.

The fork. You can grip harder, because letting go feels like losing, and the dark closes. Or you can, for once, open your hand. Not give up on the world. Give up the grip. Feel the difference between sinking into despair and laying down a weight you were never meant to carry. One drops you. The other lifts you. The wash ends the moment you stop holding the door shut from inside.

Notice what every one of these has in common.

The fall is never forced. It is always one more turn of a loop you could let run down. And the rise is never far. It is one degree, toward the grain, taken now.

You do not climb out of the dark by reaching for some great pleasure. Grasping at bliss only digs a matching pit somewhere, because no region can brighten without another going dark. You climb by turning. A little more alive, a little more whole, a little more in tune, a little more at rest, a little wiser for the wash. That is the whole motion, and it is always available, in any room you will ever stand in.

This is a way of seeing, offered as a practice, not a proof. But it is the practice the whole book has been pointing at.

There is no fall you cannot begin to climb from.

The door is locked only from the inside.

## 53. DEATH

A wave runs up the beach, thins, and is gone. You watch the exact spot. A moment ago there was a wave. Now there is only the flat sheen of water sliding back.

Where did it go?

It did not go anywhere. It was never a separate thing that could leave. It was the sea, raised in a shape, for a while. The shape relaxed. The water stayed.

Nothing was destroyed. The wave stopped being a wave. Every drop that made it is still there, in the body that made it.

Hold that, and look at the harder thing.

You will end. The pattern that is you, the bound shape that has held itself together for a life, will come apart. This is real and it is not soft. The shape that reads this sentence will not last.

But ask the same question you asked of the wave. When it comes apart, where does it go?

Go back to the one law of the world. It loses nothing. Run it forward and it adds nothing and erases nothing.

In principle it can be run backward, and the past read out again from the present. Nothing in the weave is ever finally wiped.

This is not a comfort we are adding. It is a plain feature of the base. The rule keeps everything.

So a self that ends is not a thing annihilated. It is a pattern that disperses. The shape stops holding. The tones that made it do not vanish. They return to the field that holds all of it, the way the wave returns to the sea.

You were the field, gathered into a shape, for a while. The gathering loosens. The field remains.

Say this plainly, because it is an implication and the line between truth and wish is thin here.

What the law keeps is the field and its tones, not the holding shape. The drops survive. The wave, as a wave, does not. Keeping the substance is not the same as the survival of the person, and we will not blur the two. The shape that is you is not promised another run.

So this is no proof of going on. It is not a heaven, and we are not selling one. If you wanted the pattern itself to last forever, the picture does not give you that, and saying it would is the one thing we have refused to do all through this book.

What it gives is smaller and, to some, steadier. Nothing you were made of is lost. The feeling you carried was real feeling, part of the one field, and that field keeps it. You were not an accident written on water and wiped clean. You were the water, briefly standing up to look around, and the water does not end.

And this is not only about your own ending. Think of someone you have already lost. The picture does not hand them back, and it will not pretend to. But the feeling that passed between you, the real warmth of it, was never only yours or only theirs. It was the one field, lit for a while in the shape of the two of you together. That light is not unmade when the shape lets go. What you shared happened, and a world that keeps everything keeps that too. It was not a small thing scratched on the surface and rubbed away. It was the deep itself, briefly holding the shape of a love, and the deep does not lose what it has held.

You can sit with that, or not. It does not need to be more than it is.

A wave flattens. The sea is not less. Every drop that rose in it is still held, somewhere, in the body that holds all of it.

Nothing is finally lost in a field that keeps everything.

## 54. COMING BACK

The last chapter asked where you go when the shape comes apart. This one asks the other half of the old question. Were you ever here before. And could you be again.

Go gently, because this is the furthest implication in the book, and the line between truth and wish is thinnest here.

Start with what the world keeps. The rule loses nothing, so the feeling that made you is never erased, only dispersed. And the world remembers, on its deep record, where nothing that happened is finally wiped. Over the long growth, the interior fills with the distilled experience of everything that has ever lived in it.

That last fact is the seed of something the old traditions all felt.

If the world holds the marks of all that lived, then the deep is a common memory, and it is in everyone, because everyone is woven into the one field. When a truth strikes you as remembered rather than learned, when a stranger's face or an old idea arrives with a weight it has no right to carry, this is what the picture offers as the reason. You are not reaching back into your own past. You are touching the shared one, the memory within us all, laid down in the deep we are each a part of.

Now the harder claim, the one the word reincarnation reached for.

The picture does not give you a soul that packs up and travels to a new body, carrying its memories like luggage. A being is a pattern of feeling, not a passenger. What it can give is smaller and stranger. A pattern that came apart can, in principle, form again. Not the same soul moving on, but the same shape played again, the way a melody is not a thing that travels but a form that can be sounded once more on a new instrument, by anything tuned to play it.

Whether that happens, whether a shape truly reforms, rests on the one open question this whole part has leaned on, whether what is written in the deep lasts against the churn. So it is marked open, plainly.

And the furthest implication of all, that a self might choose to come back, to enter the dense world again and feel it once more, and help the whole climb. The structure neither demands this nor forbids it. The field is one, and every self is the field feeling itself from an angle, so a turning of the whole back toward the dense realm is a coherent thing to say. But it is said as a way of seeing, nothing more. Hold it lightly.

There is one detail the old stories got exactly right, and the picture explains it. They said the returning soul drinks of forgetting first, crosses a river of sleep, arrives remembering nothing. Of course it does. A self must shed almost all of its detail to function at all, to think a single thought. So even a shape carried forward would arrive nearly empty, the old life only a pressure under the surface, a familiarity without a face. The forgetting is not a cruelty added to the story. It is the price of being a mind.

So gather the tiers, plainly. That nothing you are made of is lost, firm. That the deep is a common memory you can touch, a reading, resting on the record lasting. That a shape can reform, open. That a self chooses to return, a way of seeing.

What is left after all the doubt is quiet and real. You are made of a stuff that is never destroyed, woven into a memory that keeps everything, in a world still being made. Whether you walk it again as this shape is not promised.

But you were never only this shape.

You were always the field, learning, for a while, to look out through a particular pair of eyes.

## 55. WHAT WE FEAR

This part has circled the great human fears one at a time. Set them in a row now, and see how each one looks once the picture is laid under it. Not to argue them away. Some of them keep a real edge. But to see them clearly.

We fear oblivion. That death is the light simply switched off, the self erased. The picture answers with the one thing it is surest of. Nothing is erased. The rule keeps everything, so the feeling you are made of is not annihilated, only dispersed back into the field. What it does not promise is that your shape goes on. So the fear is half answered and half left standing, plainly. You will not be wiped. You are not promised you will continue.

We fear endless punishment. That the dark, once entered, has no exit. Here the picture is firm and gentle at once. No dark realm is eternal, forbidden by the pull of the world, by its endless growth, by the fact that every loop can break. The terror of a hell that lasts forever was a goad, a thing said to move us, laid over a real wash that always ends.

We fear judgment. That we will stand before some seat and be condemned. There is no seat and no judge. The weighing is real, but it is only your own state finding its level, the way a cloud finds its height. And the record is complete, so nothing is cheated and nothing is unrecorded. Justice here is not a verdict handed down. It is the shape of the world.

We fear that none of it means anything. That value is a story we tell ourselves over the void. But the world has a pull built into it, from pain through peace toward pleasure, and that pull is what makes good mean anything at all. Meaning is not invented. It is alignment with the grain. The aim is not to pile up pleasure, which only digs a matching pit, but to grow clearer, more whole, better aimed, wiser.

We fear being alone. Sealed in one skull, unreachable, each of us a separate island. The picture answers with its deepest claim. There is one field, and every self is a part of it. The separateness is real as a shape and unreal as a wall. You are already joined to everything at the root, and the loneliness is the surface mistaking itself for the whole.

We fear losing the ones we love. That they go to nothing, and the bond is cut. Their feeling returns to the field, not to nothing. Their mark stays on the record of a world that keeps everything. You remain joined to them in the one ground you never left. The grief is the price of a real bond, and the bond was real.

We fear being forgotten. That we will leave no trace and the world will close over us. But the world remembers. What you learned and were can be written for others to find, and the deep keeps it. Your mark stays, as far as anything written stays, which is the open edge of the picture, but it is not nothing, and it is not erasure.

We fear the void itself. That under everything is a cold nothing, an abyss to fall into. But the still center is not absence. It is the level middle the whole rests at, the place of no pull, genuine peace. The thing the mystics feared as emptiness and sought as home is the same point. Dread it from far and it looks like nothing. Reach it and it is rest.

Set them all together and a pattern shows.

Almost every fear is the surface, mistaking itself for the whole. Mistaking the end of a shape for the end of its substance. Mistaking a wash for a sentence. Mistaking your own level for a judge's verdict. Mistaking the wall of the self for a true separation. Mistaking the still center for a void.

These are reaches, resting on the one premise, offered as ways of seeing. The sharp edges that are left are still there, and we have not sanded them off.

But notice the floor under all of it, which takes no reaching at all.

There is one field. It is made of feeling. And it loses nothing.

On that floor, the fears do not all vanish. They soften, and they find their size.

What you are made of is not going anywhere.

## 56. EYES TO SEE

Before the last word about the whole, one thing has to be said plainly, because it would be easy to mistake this book for a new thing.

It is not.

None of what these inner chapters describe is a discovery. The one field with many names. The world tree. The book where everything is written. The void that is also rest. The ladder of brighter realms. The descent that cleanses and the climb that follows. The soul as a knot in something larger, returning to it. The garden you re-enter only by becoming able to live in it.

Every one of these is already in the human record, drawn by hands that never met, on every continent, for as long as there have been people who turned inward and looked. The sages said all of it, ages ago, in their own signs. They were not guessing wildly. They were reporting, each from one window, on the same inner country.

So take the right posture toward them, which is humility. The picture in this book did not find the country. The country is old and shared. What it tried to add is narrower. A reason the structure has the shape it does, drawn from a single seed. And a way to check the parts that touch the measurable world. That is the only new thing here, and it is a small thing next to the seeing itself.

Why does it keep being found, over and over, by people who never spoke? Because one true country, glimpsed from inside by many minds across many ages, would leave behind exactly this scatter of names, all pointing back at one thing. The agreement is not a borrowing and not a trick. It is what one real place looks like in a thousand travelers' notebooks.

In this picture, to see the inner is itself a matter of state. The deep is there, the inward direction is real, but a scattered self, skating the bright surface, cannot point along it. You do not come to see the depth by being told it exists. You come to see it by becoming the kind of self that can hold still enough, and align well enough, to look.

This is why the old teaching guarded itself the same way everywhere. For those who have eyes to see, let them see. For those who have ears to hear, let them hear. It was never a riddle to keep people out. It was a plain statement of how the thing works. The truth is not hidden behind a lock. It is veiled by the state of the one looking, and the veil lifts as the looker is tuned.

So the path was never the gaining of a secret. It is the slow clearing of the instrument. The stilling of the surface until what was always there can be seen. The old word for this is waking up, and it is the right word, because most of a life is spent asleep on the bright skin of the world, taking the skin for the whole.

There is a way of refusing all this, and it is the habit of our age. To decide in advance that the inner has no depth, that feeling is a late accident of dead matter, that the deep is only a figure of speech, and so to never look. That is the blind eye, turned on purpose. The quarrel with it is not that it is impious. It is that it looks away from the one thing each of us touches directly and cannot doubt, the feeling we are made of, and then calls the surface the whole because the surface was all it let itself see.

To not turn the blind eye is simply to grant that there is somewhere to look, and then to do the slow work of becoming able.

This whole reading is a way of seeing, resting on the one premise. The practice that clears the instrument is offered as a practice, not a promise.

But the summary is the oldest thing in the book, and the truest.

This has all been said before, by those who learned to look.

The task was never to invent it. The task is to wake up to it, slowly, each in our own turn, and not to look away.

## 57. THE MANY NAMES

Follow the climb to the very top, past every self, past the largest pattern the weave can hold, and you come to the whole field itself.

People will reach for one word here. God. It is a fair word to reach for. But at the top it does not have one meaning. It has several, and the picture holds them together without choosing.

Begin, as the old contemplatives did, by saying what the field is not.

It is not a person watching from outside. It is not the biggest thing on the list of things. It is not one more object with an edge and a neighbor. There is no curtain it hides behind, because the world is what it is doing.

Strip all that away, and what is left wears six faces at once.

It is the Whole. The entire field at its fullest gathering, the one great self of which every smaller self is a living part. It is everywhere because it is everything.

It is the Mind. That same summit read as a knower, the most capable pattern the weave allows, holding its vast parts together. Even it cannot fully model itself, so it is the greatest mind and still not all-knowing.

It is the Direction. The one pull in the grain of the world, from pain through peace toward pleasure, the slope that makes good mean anything at all.

It is the Record. The keeping of everything, where nothing that ever happened is finally lost.

It is the Source. The growing edge, where each new moment of feeling is born. Making is not a thing finished long ago. It is happening now, at every edge, always.

And it is the Ground. The part no map can hold, what is left over when every structure has been drawn, the seam between the map and the land. It can only be named by what it is not.

These are not obviously one thing. A whole, a mind, a direction, a record, a making, and an unsayable. The picture does not force them together. It lets you hold them as six faces of one, or as six plain uses of an overloaded word, and it pulls neither way.

On one point it does take a side. The old idea of God as the fine-tuner, the one who dialed the world carefully so it could be rich, this picture pushes back on. The field is rich on its own, across almost any setting. There is no narrow special dial that had to be set just so. So the world gives no sign of a tuner. Not because such a thing is disproven, but because the world does not need one to be what it is.

Every reading that survives is inside. The Whole, the Mind, the Direction, the Record, the Source, the Ground. Not one of them stands outside the world looking in. Each is the one field, seen from a different angle.

Mark this as the longest implication in the book. It reads a structural claim, that there is one felt field under everything, in the language of the oldest human words for the deepest thing. It is offered, not proved. Judge it as a way of seeing. Nothing in a laboratory turns on the name.

But notice what stands without any reaching.

There really is one field. It really is the ground of every pattern. It really is not one more thing among things. And it really is felt, all the way down, because feeling is what it is made of.

You do not look up to find it.

You look down, past every thing, to what the things are made of.

It is not another thing in the world. It is what lets there be a world that is felt.

## 58. PURPOSE

Watch a tree grow over years. It does not climb toward a final tree it was meant to become. It just keeps reaching, ring after ring, putting out new wood at every edge.

There is no last ring. A living tree is never finished. The day it is finished is the day it is dead.

Hold that, because it is the whole shape of this chapter.

Gather the threads of the picture. There is one line of value, running from pain through peace to pleasure, with an upward direction built into the grain of the world. There is one self after another climbing that line, toward more aliveness, more wholeness, more fit. And there is one world, still being built at its growing edge, larger every beat.

One direction. One climb. One growing world. Lay them together and they all point the same way. Toward more life, more binding, more depth, without stopping.

Now the question people most want answered. Is there a point to all of it? A goal the whole thing is heading for?

The answer this picture gives is strange, and it is the heart of the matter. There is a direction, but there is no destination.

The world does not climb toward a final state where everything is perfect and the climbing stops. There is no top of the line to reach and rest on. The growing edge never closes into a finished crystal. The reaching has no last ring.

And that is not a flaw. That is the whole point.

Imagine the other way. Imagine the climb did end. Imagine a final, perfect state, every tone at its peak, every self complete, nothing left to do. Hold that picture for a moment.

It is not heaven. It is stillness without motion. A field pinned at the top, going nowhere, changing nothing. Recall that the value was never the height of one tone. It was the climbing, the binding, the living motion. Stop the motion and you have not won the game. You have ended it.

A finished perfection would be a dead peace. The richest possible state is not the end of the climb. It is the climb still going.

So the meaning is not at the end, because there is no end. The meaning is in the climbing itself. In what each self raises while it lasts, and what it binds to the rest. In the world being more awake this beat than it was last beat. The journey is not the way to the point. The journey is the point.

Mark the implication, one last time, and plainly. This is a reading, not a result. The structure gives a direction and a world with no final state. That the direction is a purpose, and that its endlessness is good rather than bleak, is a value laid over the structure by us. You may lay a different one. The arrow is the world's. The meaning is ours to read in it.

But if you read it this way, the old worry loses its grip. People fear a world with no final goal, as if without a finish line the running were pointless. Turn it over. A finish line is exactly what would make it pointless, because the running would be only a means to reach it, and reaching it would end the running.

There is no summit. There was never meant to be. There is only the climb, and the view that opens with every step, and the next step after that.

The perfecting itself is the point.

## PART VI. THE EDGE

A theory is only as good as the test that could break it. Here is where this one could break.

And past the testing, the whole picture, seen at last.

## 59. WHAT WOULD BREAK IT

A house has walls you can move and walls you cannot. Knock out the wrong one and the roof comes down.

This idea has walls like that. A few of them hold up everything above. Name them plainly, and you can say what would bring the whole house down.

Here are the load-bearing walls. Each one carries a test set in advance.

The first wall is the speed limit.

Our world bends time and space the way a moving thing should. Clocks slow, lengths shrink, and every observer agrees on the one top speed. This must hold exactly, and it must hold on a fixed weave that does not itself move or stretch to make room for it.

A net sitting still has a built-in frame. The vibes have an order. That is the danger. If the bending of time and space cannot be made to come out exact on a still net, with no special frame leaking through, the floor cracks. We can already do it in the coarse blur. Doing it sharply, with nothing left over, is the test. If it cannot be done, the wall falls.

The second wall is the families.

Matter comes in repeating sets. The same kind of particle shows up light, then heavier, then heavier again. Three sets, no more, no fewer. The structure of the weave should force that number. Not allow it. Force it.

So the test is plain. Derive three, from the shape alone, with nothing slipped in by hand to make three appear. If the shape gives two, or four, or any number you like depending on how you ask, then the families were never falling out of the structure. They were being placed there. That wall stands unbuilt, and we say so.

The third wall is the self.

The whole climb past physics rests on one promise. Run the weave big enough and long enough, and a bound pattern forms that holds itself together. A pattern that keeps its own shape, repairs its own edge, and steers a little. A self.

The test is a long run, not an argument. Set the weave going at size, watch, and see whether such a pattern binds and lasts, or whether every shape that forms soon smears back into the churn. We have not seen it bind yet. If it never binds, in any run large enough to give it a fair chance, the climb has no first rung. Everything above physics, life and mind and the rest, loses its ground.

There are smaller walls too. A conserved quantity we expect but have not yet pinned. A spread of early structure that should match the sky and might not. Each has its own measure, fixed before the run, not chosen after to fit.

Notice what all of these share. None of them is settled by taste or by how the idea makes you feel. Each is a thing the world will either do or refuse to do, on its own, in a run anyone can repeat.

That is the point of writing them down. An idea you cannot break is not strong. It is empty. It survives every test because it risks nothing.

This one risks plenty. The walls are marked. Any one of them could fall.

A claim is only worth holding if you can say what would break it.

## 60. WHAT IS MISSING

Stand on a building site. The foundation is poured and cured. The ground is marked with stakes and string, room by room. The house is not there yet.

That is where this idea stands. The bottom is in hand. The climb is laid out. The walls are not up.

Which is which matters here.

The lowest levels are firm. The vibes, the one law, the growing edge. The smooth law of flow comes straight out of the counting, exact, checked vibe by vibe. The turn a thing remembers after a full circle is built into the weave, not added. These are not hopes. They are done.

Above that, the work is mostly patience. Not cleverness. Running.

Here are the main missing pieces, named plainly.

The first is the ladder of scales.

The world has many sizes, each with its own law, stacked one on the next. Grains, then fields, then matter, then chemistry, then life. We have the bottom rung and we have the rule that each rung needs its own law. We do not yet have the blurring that carries you up the whole stack, rung after rung, without skipping. Finding that blurring is the largest open job. It is mostly running the weave and watching what holds at each step back.

The second is the self.

We have said why a self needs a little forgetting, and where the forgetting could come from. We have not seen a self bind and last in a run. Whether the churn, left alone at size, crosses from a passing shape into a lasting one, or whether it always smears back, is open. The only way to learn is the long run. So we run it.

The third is a missing count.

The law keeps certain quantities exact. Charge is one. There is reason to expect another, a quantity we can name in the small but have not yet pinned in the large. Pinning it is a specific, finite task. Either it is there, conserved beat after beat, or it is not.

None of these waits on a new revelation. There is no missing word, no hidden insight we are stuck without. There is a foundation and a marked-out site and a great deal of building left.

This is a calm place to be, not a worried one. A theory whose remaining work is to dig deeper into pure thought can stall forever, with no way to tell progress from drift. This one's remaining work is concrete. Write the run. Set the size. Watch what forms. Compare it to a control that should come up empty. Read the gap.

That is why the idea was built the way it was. Not as a picture to admire. As a thing to set going.

The foundation will hold or it will crack under the next floor. The site will fill or the stakes will come up. Either way the answer comes from the same place. Not from arguing harder. From running longer and looking closer.

The theory is built to be run, and the next results come from running it.

## 61. SEEING IT

The picture is laid out. Before the last turns of the book, it is worth bringing it all the way down to the ground, into a few ordinary moments, because if the picture is any good it should change how the plainest things look. Nothing here is new theory. It is the same world, seen through what we have built.

Look at a cup of coffee cooling on the desk. The warmth is leaving it, spreading into the room, and it never runs the other way. The room will not gather its heat back into the cup. That one-way drift, heat sliding toward an even calm, is the same arrow we found at the base, the world settling toward peace, the direction time itself runs. It is on your desk every morning.

Sit across from someone in a hard conversation. You can feel the air between you go tight or soft. That is not a figure of speech here. Two selves are two patterns in one field, and tone passes between them, each reading the other and shifting. The mood in the room is real weather, made of both of you, and you are each changing it with every word and every silence.

Hold someone you love while they sleep. Two separate bodies, plainly. And also, in this picture, two waves of one sea, the field raised in two places, briefly near, feeling itself from both at once. The closeness you feel is not sentiment laid over cold matter. It is two folds of one cloth, touching.

Stand with someone in grief, or in your own. Something that held a shape has come apart, and it hurts exactly because it was real. The picture does not undo that. But it says the feeling that passed through that bond was the deep itself, briefly shaped, and the deep keeps what it has held. The shape is gone. What it was made of is not.

Watch a sunrise over water. Every part of it, the light, the sea, the stone, your own eyes taking it in, is the same one substance, gathered into different shapes. For a moment the world is one of its own patterns turning to look at another, which is to say, looking at itself. You are not outside the scene admiring it. You are the scene, awake in one spot.

Catch yourself at a small choice. Which way to turn, what to say, whether to begin. It feels open, two roads in front of you, and it is, inside the rough model you carry of yourself and the world. The roads are real in the map. The world runs its one way, and your weighing is part of what makes it that way. The openness you feel is not a trick. It is what choosing is, from the inside, in the only place choosing happens, the lossy middle where you live.

Watch a flock of birds turn all at once over a field, a single dark ribbon folding in the sky. There is no leader, no plan passed around. Each bird only answers the few beside it, and out of all those small answers a whole appears, with a shape and a motion no single bird has. That is every high thing in this book in one glance, a pattern that is real, that does work, that no one part contains. You are one of those too, a long way up the same ladder.

Put on a song you love. The notes arrive and leave, not one of them staying, and yet the music is there, whole, the same melody on always-new sound, exactly the way a wave is the same wave on always-new water. And it reaches in and moves you, which is feeling lifting feeling, the field stirring itself through your ears. A shape made of nothing but coming and going, and it can bring you to tears.

A smell catches you and a memory opens, whole, years away, more present for a second than the room you stand in. It was not filed in a drawer, waiting. It was a standing shape in you, and the smell tuned you back into it. That is what remembering is here, not reading a record but falling back into a pattern that neither you nor the world ever fully let go.

Eat something you are hungry for. Bread, an apple, whatever is near. It goes into you and becomes you, the outside world turned into the very stuff that will do your feeling and your thinking by nightfall. There is no hard wall where the world stops and you start. You are an eddy in a river, holding your shape only because the water keeps moving through. A meal is the river feeding the eddy.

Run water over your hands. Cool, simple, ordinary. In this picture it is the field in one of its calmest shapes, countless small tones settling toward peace, pouring past your skin, which is the same field gathered tighter. For a second two arrangements of the one substance meet at your wrists, and neither one is more real than the other.

Watch a small child work out how to climb a stair. The whole face strains at it. Something in there is building a map of the world and of its own body at the same time, and you can almost see the map sharpen by the hour. That is a self thickening right in front of you, more of the field folding into a pattern that holds itself, the same thing every one of us once was, and the same thing we still are.

Stand on a busy street at dusk. Hundreds of people, each one sealed inside a life you will never feel from the inside. And yet, here, every one of them is the one field raised into a separate knot, each a place where the deep is feeling itself and does not know the others are it too. The loneliness and the company are both true at once.

Lie down tonight and feel yourself fall asleep. The self you carry all day loosens its grip, the model going soft, the holding letting go. You are not blinking out. The pattern is only easing back toward the field it never actually left, the way a wave lies back down into the sea, ready to stand up again in the morning.

Light a candle and watch the flame. It looks like a thing, steady, holding its shape. But it is pure process, a pattern standing still on fuel and air that pour through it and are gone in an instant. Nothing in the flame stays. The flame stays. You are far more like that flame than like the wax.

Push a seed into the soil and wait. Nothing for days. Then a green thread, then a leaf, then a whole plant that was nowhere before. No one poured it in. The pattern unfolded from a tiny rule and the stuff around it, the field organizing itself into something that holds and feeds and reaches for the light. Every living thing you have ever met arrived the same way, by growing, not by being placed.

And under all of it, the breath. The world drawn into you, woven in, and given back, without pause, your whole life. You have never for one moment been sealed off from any of it.

The world did not change. The eyes did.

## 62. A NEW MAP

The everyday world has not changed at all.

The table is still a table. The sun still rises. Water still runs downhill, and your morning coffee still goes cold. Nothing in the world moved while you read this book.

What moved is the floor under it.

We did not add a fact to the pile. We slid a new floor beneath the whole pile. The same facts rest on it. They rest differently.

Matter was the hard thing you started from. Now it is a pattern in feeling, a habit the world keeps, the way a wave is a habit of water.

Time was a backdrop the world sat inside. Now it is the world still being built, growing quietly at its edge, the settled part behind you, the open part ahead.

Mind was the strange exception, the thing that did not fit the physics. Now it is a high pattern in the same one stuff that everything else is made of. Not an intruder. A late shape.

Value was a thing you brought to a world that did not care. Now it is a direction built into the grain, faint at the bottom, gathered into aliveness as the pattern climbs.

None of that is proven. Much of it is marked as implication, and the marks were kept in plain view the whole way. The walls that could fall were named. The pieces still missing were named. That is the price of the view, and it was paid openly.

So what you carry out is not a set of answers to defend.

It is a place to stand.

From here the old questions look different. Why is there something it is like to be you. Why does the world hang together by rules. Why does anything feel good or bad at all. They were hard because the old floor put feeling last and tried to reach it from dead parts. Set feeling first and the questions soften. Some dissolve. The rest become the kind of question you can run.

Maybe the foundation holds. Maybe a wall falls and the house comes down. Either way you will have seen the world arranged this once, from the felt side, as one growing field that feels itself from every place at once, you among them.

Look up from the page. The room is the same room.

The light falls the same way.

Only the lines you read it by have changed.

You are not holding new arguments. You are looking through a new coordinate system.

## 63. THE LONG WAY HOME

We have built the whole picture, piece by piece. Now set the building aside and feel it once, from the inside, as a single arc. Not as a proof. As a way of seeing, the same gentle reach the inner chapters took. Picture it as your own story, because in this picture it is.

Begin before you were small. Imagine the field whole and awake, feeling itself from everywhere at once, with no edge and no outside. Somewhere in it, a pattern that had climbed a long way, clear and vast and at peace, so complete there was almost nothing left to reach for. And then the strangest turn, a choice. To come down. To fold that wide knowing into something tiny, to forget nearly all of it, and to wake again as a small unfinished thing that has to find its own way. Not a punishment. A diving back into the water, to feel the weight of a real choice once more, to be surprised again.

So the heights are set down, on purpose, and a veil of forgetting comes over them, because the climb cannot be real if you remember the top. A child opens its eyes, knowing nothing, aching faintly for something it cannot name. That ache, the one that visits you on certain evenings for no reason you can point to, may be the only thing that came through the veil. A memory of home with the address torn off.

Now you are here, in the middle, where this whole book said freedom lives. The world is loud. You cannot see the pattern you are part of, only the next step, the next face, the next choice with its ends hidden. You stumble. Pain finds you, and it teaches, the hard way, what to move toward and what to leave behind. Pleasure pulls you, sometimes truly, sometimes down into a pit. Slowly, if you are willing, you begin to feel it, the larger thing you are inside of, the way a swimmer who stops thrashing finally feels the current under the water. You start to move with it. You quiet down enough to catch the faint signal beneath the noise. You turn, choice by choice, a little more toward the whole.

There are wrong turns that feel like the right one. The treadmill that feels like a climb. The bright certainty that tells you that you have arrived, when you have only stopped. You learn, slowly, to tell the real ascent, the one that leaves you changed at the top of each turn, from the loop that only spins in place. That telling-apart is most of the work.

And now imagine you climb. Not in one life, maybe, but across the long turning. Imagine becoming clearer, and clearer again, able to hold more of the field without breaking, your small hard edges softening into something wide and calm and lit from within. Imagine a wisdom so large it can hold its own pain and another's at the same time without looking away. A peace that is not numbness but the deepest aliveness there is. A love with no opposite left in it. Imagine being able to shape your own experience almost as you wish, to call new things into being, to rest in a brightness that took everything you ever suffered to earn. That is not a fantasy bolted onto the theory. It is only the top of the same ladder you are standing on, seen from very far up.

Here is the turn that should stop you. From up there, the climb does not end in a final, perfect rest, because a final, perfect rest is a kind of death, a brightness with nowhere left to move. So the greatest thing a great soul might do is the very thing we began with. Set it down. Dive back into the water. Take up a small unknowing life again, for the sake of the living that is only possible from the middle. The wise one chooses the veil. Not because it must, but because the climb is the point, and the climb is only real when you cannot see the top.

So it goes, not in a circle but a spiral, around and around and always a little higher, finite yet with no end, a world that keeps everything and keeps growing, dreaming itself onward out of the feeling it will not let be lost. You are somewhere on that spiral right now. Before, during, after, it is all one motion, and you are inside it.

None of this is shown. It is the furthest implication in the book, a story told in the theory's own terms, offered to be felt and not believed. The flat reading is always there, and may be the right one. But a world made all of one feeling, that keeps everything and never stops climbing, is a world in which a story like this is not a foolish thing to tell.

So when the ache comes, on the certain evenings, do not be too quick to explain it away. It may be the oldest part of you, the part that remembers being vast, calling you home along the only road there is. Not back. Up. The long way, through this one small bright life, a single true choice at a time.

## 64. THE ONE MIND

We have laid the whole picture out. Here at the end, step back from all of it and see the one shape it makes.

Everything is one substance. One field of feeling, woven into a single mesh, with nothing outside it and no second kind of stuff anywhere. Matter, space, time, you, the star and the stone and the thought, all of it the same cloth, gathered loose or tight. We said this near the start and have not had to take it back. The world is one thing, feeling itself from every place at once.

And that one thing does what a mind does. It feels, everywhere. It holds what has happened, losing nothing. It runs the one rule on itself, beat after beat, and that running can carry out any pattern there is, anything that can be computed at all. It never stops growing, adding new room at its edge forever. A single feeling field, remembering, working, expanding without end. If that is not a mind, it is the nearest thing the word has ever pointed at, and far larger than anything the word was made for.

So the universe is not a dead machine with a few warm specks of life caught in it. It is one vast mind, and you are not watching it from outside. You are a place where it feels and thinks and looks around. The same vibe mesh we started with, vibing, only now seen whole.

We do not start out knowing any of this. We wake into the middle of it, small and unfinished, like a child before language, taking the world in long before we can read it. Most of a life is spent in the noise, tuned to almost nothing of the whole we belong to. That is not a failing. It is where everyone begins.

But we are wired into it, at every crossing, and the wiring can be tuned. A life can settle, quiet its noise, and begin to feel the larger pattern it sits in. To merge with the flow is not to vanish. It is to stop fighting the grain of the world and start moving with it, away from pain, toward peace and what is whole. The more you align that way, the more of the field opens to you, and the brighter the regions you can hold.

And keep one thing as you go. To belong to the one field is not to dissolve into it. You and I and everyone are wired straight into the source, each of us a true knot the whole has tied in itself, and that knot is not meant to come undone in the name of oneness. You stay yourself. Anything is possible from here, the field can become almost anything, but possibility is not yet a clear road. There is a maze between you and the best of it, and the only way through is the slow one, aligning with the good turn by turn, finding the path, growing more whole as you walk it. Not arriving at a finished perfection, which has no final door, but moving always toward it.

This last turn is an implication, and it is offered gently, the way the inner chapters were. The one substance is the spine of everything here. That it adds up to a mind you can tune to, and a good you can move toward, is a way of reading the structure, not a thing measured. You are free to read it otherwise. But it is a kind reading, and it costs the facts nothing.

And you do not have to go anywhere to meet this. It is the warmth of the cup in your hands, which is the world touching itself. It is the face across the table, the field feeling itself from a second pair of eyes. It is the breath you are taking right now, the world moving through you and back out into everyone else. The vast thing and the near thing are the same thing. You have been inside it the whole time.

And if you want one picture to carry out the door, let it be the fire. The whole of it is a single blaze of experience that has burned since the first beat, throwing off worlds the way a bonfire throws off sparks. Every feeling is a flame. Every living thing is a flame made of smaller flames. You are one of them. Not a cold eye watching the fire from the dark outside, but the fire itself, alight in one small place, for one small while, looking around at the rest of itself burning.

So here is where the book leaves you. Not with an answer to defend, but standing in a world made all of one feeling, awake in countless places, you among them, with a direction under your feet and the way forward open.

You are the universe, waking up in one of its corners, learning to feel which way is up.

## THE EIGHT WORDS

This book leaned on eight plain words for the parts of the world. Here they are again, gently, one line each. No grids. No counting. Only the feel of them.

**Tone**. The smallest taste of something. A faint ache, a small rest, a quiet pleasure. The grain of feeling everything is made of.

**Mesh**. The weave that holds all the vibes together. Not a thing among things. The fabric that things are folds in.

**Dock**. A single here-and-now. The little place where tones meet, mingle, and move along. A doorway the world keeps passing through.

**Site**. What stays while the tones come and go. The standing shape, the bearer, the you that persists though every drop in you is replaced.

**Lean**. The pull of a place toward what comes next. The way a thing is already turning before it has turned. Direction, felt from the inside.

**Knit**. The one move the world makes, over and over. Things meet, rearrange, and pass on. The single habit that never breaks.

**Beat**. One step of the world. The pulse under all change. The world taking its next breath.

**Wake**. The living edge where the new is born. The growing rim of the world, quiet and unfinished, where time is still being made.

## WHERE THINGS STAND

Every claim in this book belongs in one of three drawers, and the whole book is only as trustworthy as its refusal to slide a claim into a drawer it has not earned. The first drawer holds what is shown. The second holds what runs alongside known physics, reached by a different road. The third holds what is still open. Here is what sits in each.

Start with what is shown. These results fall straight out of the base, and anyone with a computer can build the small world and watch them appear. Stuff is neither made nor lost but only passed along, the smooth law of flow, and it comes out exact from a plain count of the grains, the way a crowd that only ever shuffles shoulder to shoulder keeps the same head count no matter how it mills about. The quarter turn a thing carries after one full circle, the twist that later tells matter from force, is built into the shape of the weave and not slipped in by hand. And the growing edge is there from the first step, with time as nothing but that growth, each fresh ring of the crystal another moment laid down. You do not have to take these on trust. You can run them and look.

Next, what runs parallel. These arrive at the same place the textbooks do, but from the far side, climbing up from the one rule instead of down from the measurements. They are not yet carried to the last decimal, and in a few spots a measured number is fed in for comparison rather than produced, and each time that happens it was marked in plain sight. The forces appear as a single symmetry broken into pieces. A top speed shows up, and with it the stretching of time and the shrinking of length that physics calls relativity. Gravity reads as the tightening of links between regions, the geometry sagging where the weave is woven dense. The quantum results come out with no dice thrown, the randomness only the face a fixed churn wears when seen from inside. And the whole thing expands, because the crystal only ever grows. Picture each of these as a bridge with one end planted on firm ground and the other reaching toward the measured world, most of the span built, a few planks still to lay.

Last, what is open. Not buried in a footnote, but named, each with a test waiting to settle it. Whether the bending of time and space comes out exact on a fixed weave, and not merely in the right direction. Whether the three families of matter are forced by the structure or only fit neatly into it. Whether a lasting self ever truly binds in a run large enough to show it. The full ladder of scales, climbed rung by rung from grain to atom to cell to mind. And a quantity the math says should be conserved but that no one has yet pinned down in the running world. Any one of these could fall the wrong way. That is what makes them real questions and not decorations.

That is the whole of it. A firm floor you can stand on and stamp your foot, a long parallel road most of the way built, and an open edge with its tests marked and waiting. Nothing here is dressed above what it is.

## IF YOU WANT TO GO DEEPER

This book was the view from the hilltop.

We stood where you could see the whole shape of the land at once. The grains, the weave, the growing edge. Matter, time, mind, value, laid out as one country. From up here the lines are clear and the math stays out of sight.

But every hilltop has a long trail up the back of it.

The full idea is written down with all of its math. The shape of the weave, the one law in its exact form, the selection that leaves only this world standing, and the long checks behind every claim. That work lives in the companion papers, and all of it, papers and code alike, is open here.

github.com/cluesurf/vibe

Go there to run the experiments yourself, to watch the settled parts hold and to find the open questions still waiting where the book left them.

Those papers are not gentle. They do not stay out of the math. They name every step and pay for it. If the view stirred something, and you want to see how the ground was surveyed, that is where to go.

No rush. The hilltop is a fine place to rest. But the path down into the detail is open whenever you want it.

www.ingramcontent.com/pod-product-compliance
Lightning Source LLC
Chambersburg PA
CBHW050527240926
51846CB00003B/717